I0015458

iOS Forensics Cookbook

Over 20 recipes that will enable you to handle and extract
data from iOS devices for forensics

Bhanu Birani

Mayank Birani

BIRMINGHAM - MUMBAI

iOS Forensics Cookbook

Copyright © 2016 Packt Publishing

All rights reserved. No part of this book may be reproduced, stored in a retrieval system, or transmitted in any form or by any means, without the prior written permission of the publisher, except in the case of brief quotations embedded in critical articles or reviews.

Every effort has been made in the preparation of this book to ensure the accuracy of the information presented. However, the information contained in this book is sold without warranty, either express or implied. Neither the authors, nor Packt Publishing, and its dealers and distributors will be held liable for any damages caused or alleged to be caused directly or indirectly by this book.

Packt Publishing has endeavored to provide trademark information about all of the companies and products mentioned in this book by the appropriate use of capitals. However, Packt Publishing cannot guarantee the accuracy of this information.

First published: January 2016

Production reference: 1220116

Published by Packt Publishing Ltd.
Livery Place
35 Livery Street
Birmingham B3 2PB, UK.

ISBN 978-1-78398-846-4

www.packtpub.com

Credits

Authors
Bhanu Birani

Mayank Birani

Reviewer
Dr. Aswami Ariffin

Commissioning Editor
Julian Ursell

Acquisition Editors
Vivek Anantharaman

Chaitanya Nair

Content Development Editor
Siddhesh Salvi

Technical Editor
Menza Mathew

Copy Editor
Merilyn Pereira

Project Coordinator
Nidhi Joshi

Proofreader
Safis Editing

Indexer
Tejal Daruwale Soni

Production Coordinator
Melwyn Dsa

Cover Work
Melwyn Dsa

About the Authors

Bhanu Birani has more than 7 years of experience in the software industry. He is passionate about architecting, designing, and developing complicated applications. He specializes in creating web, backend as a service, and mobile products suitable for B2B and B2C context. He has expertise in end to end development to create innovative and engaging applications for mobile devices.

After years of programming experience in different programming languages, he started developing applications for iOS devices. He started software development around the same time as his graduation and was really interested in learning about the new technologies emerging in the market. He then joined a game development company. After contributing to the gaming domain, he started working on content-based applications and radio applications. He also contributed to hyperlocal geo-targeting using BLE (iBeacons). Over the years, he has gained experience in all phases of software development as requirement gathering, feasibility analysis, architecture design, coding and debugging, quality improvement, deployment, and maintenance.

Mayank Birani has more than 4 years of experience in the software industry. He was a star from the beginning of his career, and he worked with many start-ups. Soon after graduation, he started working on iOS/Mac technologies and is an R&D engineer for Apple Inc. India. He has an inherent passion for coding and developing applications to make the world a better place. He has contributed actively to many start-ups with revolutionary ideas.

In the early stages of his career, he was the sole author of *Learning iOS 8 for Enterprise* by *Packt Publishing*.

www.PacktPub.com

Support files, eBooks, discount offers, and more

For support files and downloads related to your book, please visit www.PacktPub.com.

Did you know that Packt offers eBook versions of every book published, with PDF and ePub files available? You can upgrade to the eBook version at www.PacktPub.com and as a print book customer, you are entitled to a discount on the eBook copy. Get in touch with us at service@packtpub.com for more details.

At www.PacktPub.com, you can also read a collection of free technical articles, sign up for a range of free newsletters and receive exclusive discounts and offers on Packt books and eBooks.

https://www2.packtpub.com/books/subscription/packtlib

Do you need instant solutions to your IT questions? PacktLib is Packt's online digital book library. Here, you can search, access, and read Packt's entire library of books.

Why Subscribe?

- ▶ Fully searchable across every book published by Packt
- ▶ Copy and paste, print, and bookmark content
- ▶ On demand and accessible via a web browser

Free Access for Packt account holders

If you have an account with Packt at www.PacktPub.com, you can use this to access PacktLib today and view nine entirely free books. Simply use your login credentials for immediate access.

Table of Contents

Preface

This book focuses on the various techniques for acquisition, identification, and forensic analysis of iOS devices. This is a step-by-step practical guide that will help you to follow the procedure and extract data from iOS devices. This book helps professionals to investigate and understand forensic scenarios easily. This is a practical guide written after the rising popularity of iOS devices and the growing investigation requirements. This book deals with the various ways to investigate devices with different iOS versions and the presence and absence of other security systems such as lock code, backup passwords, and so on.

Conceptually, this book can be divided into three sections. The first section deals with the understanding of how data is generated by applications and how and where it is stored on the device. The second section focuses mainly on various analytic techniques, which include the analytics of apps by reading their logs and reports provided by Apple and other third-party vendors. This also includes the analysis and other related data mining studies provided by Google and Crashlytics. The third section, that is, the last section of the book, deals with a study, in detail, of various core forensics, which include the data structure and organization of files. This also includes the study of various open source tools that allow the detail decrypting techniques performed on any iOS device.

What this book covers

Chapter 1, *Saving and Extracting Data*, focuses on the ways to save the data in the document directory of the device along with ways to fetch the data back. In this chapter, we will also discuss the encryption and decryption of the files that are saved in the document directories. We will also discuss keychain and raw disk decryption in the chapter.

Chapter 2, *Social Media Integration*, teaches you the various ways to integrate social media with your application. The primary focus will be on the various configurations that have to be taken care of while integrating social media with the application.

Chapter 3, Integrating Data Analytics, discusses the various application analytics tools. The primary focus will be on the various ways to grab the user events and actions on the app. Then, moving forward, we will learn the powerful ways to crunch the useful information from the data gathered on the cloud.

Chapter 4, App Distribution and Crash Reporting, teaches you about the power of distributing your apps without releasing them to Apple's App Store. In this chapter, you will learn about over the air application distribution. You will also learn the various ways to track errors in our app.

Chapter 5, Demystifying Crash Reports, demystifies the overall crash reporting system. This also includes ways to read the crash reports accumulated on iTunes Connect. You will learn to decode binary crash reports to generate and gather useful information.

Chapter 6, Forensics Recovery, holds all the recovery and backup related operations on the device. You will learn about recovery modes in detail along with data extraction from the device. This chapter will also allow the user to extract data from iTunes backup files by decrypting them. Then, finally, we will see some case studies to demonstrate how data was recovered from the device in very vital cases.

Chapter 7, Forensics Tools, explores all the aforementioned tools and their capacity to retrieve data. You will learn about the various features provided by each of them. We will also compare a few tools with each other, considering both open source and paid tools.

What you need for this book

This book is designed to allow you to use different operating platforms (Windows, Mac, and Linux) through freeware, open source software, and commercial software. For a few sections, you will need Xcode 6.0 or above with a Mac OS X. Many of the other examples shown can be replicated using either the software tested by the authors or equivalent solutions and tools for iOS Forensics. Some specific cases require the use of commercial platforms, and among those, we preferred the platforms that we use in our daily work as forensic analysts. In any case, we were inspired by the principles of ease of use, completeness of information extracted, and the correctness of the presentation of the results by the software. This book is not meant to be a form of advertising for the aforementioned software in any way, and we encourage you to repeat the tests carried out on one operating platform on other platforms and software applications as well.

Who this book is for

This book is intended mainly for a technical audience, and, more specifically, for forensic analysts (or digital investigators) who need to acquire and analyze information from mobile devices running iOS.

This book is also useful for computer security experts and penetration testers because it addresses some issues that definitely must be taken into consideration before the deployment of this type of mobile device in business environments or situations where data security is a necessary condition. Finally, this book could also be of interest to developers of mobile applications, and they can learn what data is stored on these devices, and where the application is used. Thus, they will be able to improve security.

Sections

In this book, you will find several headings that appear frequently (Getting ready, How to do it, How it works, There's more, and See also).

To give clear instructions on how to complete a recipe, we use these sections as follows:

Getting ready

This section tells you what to expect in the recipe, and describes how to set up any software or any preliminary settings required for the recipe.

How to do it...

This section contains the steps required to follow the recipe.

How it works...

This section usually consists of a detailed explanation of what happened in the previous section.

There's more...

This section consists of additional information about the recipe in order to make you more knowledgeable about the recipe.

See also

This section provides helpful links to other useful information for the recipe.

Conventions

In this book, you will find a number of text styles that distinguish between different kinds of information. Here are some examples of these styles and an explanation of their meaning.

Code words in text, database table names, folder names, filenames, file extensions, pathnames, dummy URLs, user input, and Twitter handles are shown as follows: "In the popup, provide the product name as `DocumentDirectoriesSample`."

A block of code is set as follows:

```
NSArray *paths = NSSearchPathForDirectoriesInDomains(
    NSDocumentDirectory, NSUserDomainMask, YES);
NSString *documentsDirectory = [paths objectAtIndex:0];
NSLog(@"%@", documentsDirectory);
```

When we wish to draw your attention to a particular part of a code block, the relevant lines or items are set in bold:

```
NSArray *paths = NSSearchPathForDirectoriesInDomains(
    NSDocumentDirectory, NSUserDomainMask, YES);
NSString *documentsDirectory = [paths objectAtIndex:0];
NSLog(@"%@", documentsDirectory);
```

New terms and **important words** are shown in bold. Words that you see on the screen, for example, in menus or dialog boxes, appear in the text like this: "Open Xcode and go to **File | New | File**, then navigate to **iOS | Application | Single View Application**."

> Warnings or important notes appear in a box like this.

> Tips and tricks appear like this.

Reader feedback

Feedback from our readers is always welcome. Let us know what you think about this book— what you liked or disliked. Reader feedback is important for us as it helps us develop titles that you will really get the most out of.

To send us general feedback, simply e-mail `feedback@packtpub.com`, and mention the book's title in the subject of your message.

If there is a topic that you have expertise in and you are interested in either writing or contributing to a book, see our author guide at `www.packtpub.com/authors`.

Customer support

Now that you are the proud owner of a Packt book, we have a number of things to help you to get the most from your purchase.

Downloading the example code

You can download the example code files from your account at `http://www.packtpub.com` for all the Packt Publishing books you have purchased. If you purchased this book elsewhere, you can visit `http://www.packtpub.com/support` and register to have the files e-mailed directly to you.

Errata

Although we have taken every care to ensure the accuracy of our content, mistakes do happen. If you find a mistake in one of our books—maybe a mistake in the text or the code—we would be grateful if you could report this to us. By doing so, you can save other readers from frustration and help us improve subsequent versions of this book. If you find any errata, please report them by visiting `http://www.packtpub.com/submit-errata`, selecting your book, clicking on the **Errata Submission Form** link, and entering the details of your errata. Once your errata are verified, your submission will be accepted and the errata will be uploaded to our website or added to any list of existing errata under the Errata section of that title.

To view the previously submitted errata, go to `https://www.packtpub.com/books/content/support` and enter the name of the book in the search field. The required information will appear under the **Errata** section.

Piracy

Piracy of copyrighted material on the Internet is an ongoing problem across all media. At Packt, we take the protection of our copyright and licenses very seriously. If you come across any illegal copies of our works in any form on the Internet, please provide us with the location address or website name immediately so that we can pursue a remedy.

Please contact us at `copyright@packtpub.com` with a link to the suspected pirated material.

We appreciate your help in protecting our authors and our ability to bring you valuable content.

Questions

If you have a problem with any aspect of this book, you can contact us at questions@ packtpub.com, and we will do our best to address the problem.

Saving and Extracting Data

Since their launch, iOS devices have always attracted developers in ever-increasing numbers. There are numerous applications for iOS devices available in the market. While developing applications, we frequently need to save application data into the device's local memory.

In this chapter, you will learn various ways to read and write data to iOS device directory.

In this chapter, we will cover:

- The Documents directory
- Saving data using the RAW file
- Saving data in the SQLite database
- Learning about Core data

The Documents directory

Our app only runs in a "sandboxed" environment. This means that it can only access files and directories within its own contents, for example, Documents and Library. Every app has its own document directory from which it can read and write data as and when needed. This Documents directory allows you to store files and subdirectories created by your app. Now, we will create a sample project to understand the Document directories in much more depth.

Getting ready

Open Xcode and go to **File** | **New** | **File** and then navigate to **iOS** | **Application** | **Single View Application**. In the popup, provide the product name `DocumentDirectoriesSample`.

How to do it...

Perform the following steps to explore how to retrieve the path of document directories:

1. First, we will find out where in simulators and iPhones our document directories are present. To find the path, we need to write some code in our `viewDidLoad` method. Add the following line of code in `viewDidLoad`:

    ```
    NSArray *paths = NSSearchPathForDirectoriesInDomains(
      NSDocumentDirectory, NSUserDomainMask, YES);
    NSString *documentsDirectory = [paths objectAtIndex:0];
    NSLog(@"%@", documentsDirectory);
    ```

    In the preceding code, first we fetch the existing path in our path's array. Now, we will take the first object of our path array in a string that means our string contains one path for our directory. This code will print the path for document directory of the simulator or device wherever you are running the code.

2. To create the folder in `documentsDirectory`, run the following code:

    ```
    NSString *dataPath = [documentsDirectory
      stringByAppendingPathComponent:@"/MyFolder"];
    if (![[NSFileManager defaultManager]
      fileExistsAtPath:dataPath])
      [[NSFileManager defaultManager]
        createDirectoryAtPath:dataPath
          withIntermediateDirectories:NO attributes:nil
            error:nil];
    ```

    In the preceding code snippet, the `[documentsDirectory stringByAppendingPathComponent:@"/MyFolder"]` line will create our folder in the `Document` directory and the last code of NSFileManager will check whether that `dataPath` exists or not; if it does not exist, then it will create a new path.

3. Now, compile and run the project; you should be able to see the path of `Document` directories in your project console. This should look like the following screenshot:

    ```
    2015-02-16 12:49:14.277 DocumentDirectoryExample[1869:53391] /
    Users/rashmi/Library/Developer/CoreSimulator/Devices/09B756F4-
    BF50-40AE-97EE-E3FF666F4BBF/data/Containers/Data/Application/
    99A31519-F7E2-4FB7-915B-6FE9FF533983/Documents
    ```

4. Now, go to **Finder**; from the options, select **Go to Folder** and paste the `documentsDirectory` path from the console. This will navigate you to the **Documents** directory of the simulator. The **Documents** directory will look like the following screenshot:

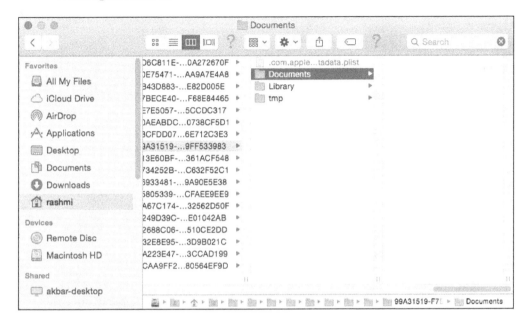

See also

The Apple developer link for documents directories is as follows:

```
https://developer.apple.com/library/mac/documentation/FileManagement/
Conceptual/FileSystemProgrammingGuide/AccessingFilesandDirectories/
AccessingFilesandDirectories.html#//apple_ref/doc/uid/TP40010672-CH3-
SW1
```

Saving data using the RAW file

In the previous section, you learned about the `Document` directories. Now, in this section we will explore this in more detail. You will learn about saving RAW files in our application's document directory.

Getting ready

We will use the preceding project for this section as well. So, open the project and follow the next section.

How to do it...

Perform the following steps to write data into your `Documents` directory:

1. Now, we will create a `.txt` file in our folder (which we created in the document directory). Customize the code in `viewDidLoad:` as follows:

```
dataPath = [[NSSearchPathForDirectoriesInDomains(
    NSDocumentDirectory, NSUserDomainMask, YES) firstObject]
        stringByAppendingPathComponent:@"myfile.txt"];
NSLog(@"%@", dataPath);
```

   The preceding code will create a `.txt` file of name `myfile.txt`.

2. To write something into the file, add the following code:

```
NSError *error;
NSString *stringToWrite = @"1\n2\n3\n4";
dataPath = [[NSSearchPathForDirectoriesInDomains(NSDocumentDirecto
ry, NSUserDomainMask, YES) firstObject] stringByAppendingPathCompo
nent:@"myfile.txt"];
[stringToWrite writeToFile:dataPath atomically:YES
encoding:NSUTF8StringEncoding error:&error];
```

   The `writeToFile` method is used to write data in our file.

3. Now, to read the data from the file, there's the `stringWithContentsOfFile` method. So add the following method inside the code:

```
NSString *str = [NSString stringWithContentsOfFile:dataPath
    encoding:NSUTF8StringEncoding error:&error];
NSLog(@"%@", str);
```

4. To see the `mayflies.txt` file in the `Documents` directory, go to the `GoToFolder` finder and paste `dataPath` from the console. It will take you to `mayflies.txt`. The final file should look something like the following screenshot:

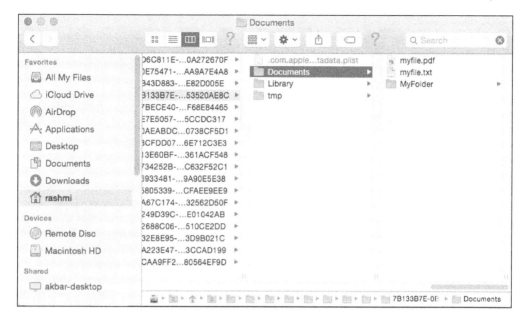

5. In the screenshot, you can see the two files have been created. As per our code, we have created `myfile.txt` and `myfile.pdf`. We have also created `MyFolder` in the same `Documents` folder of the app.

Saving data in the SQLite database

When we come across iPhone app design at the enterprise level, it will become very important to save data internally in some storage system. Saving data in the `Document` directory will not serve the purpose where there is a huge amount of data to save and update. In order to provide such features in iOS devices, we can use the SQLite database to store our data inside the app. In this section of our chapter, our primary focus will be on ways to read and write data in our SQLite database. We will start by saving some data in the database and will then move forward by implementing some search-related queries in the database to enable the user to search the data saved in the database.

Getting ready

To develop a mini app using SQLite database, we need to start by creating a new project by performing the following steps:

1. Open Xcode and go to **File | New | File**, and then, navigate to **iOS | Application | Single View Application**. In the popup, provide the product name SQLite Sample. It should look like the following screenshot:

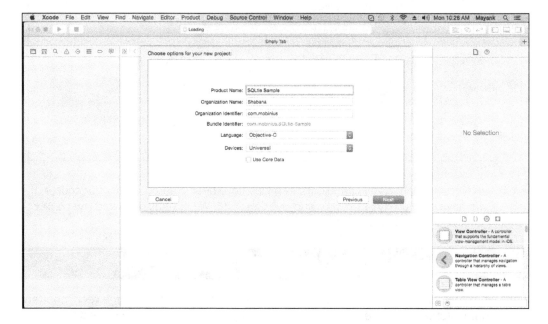

2. Click on **Next** and save the project. After creating the project, we need to add a SQLite library (libsqlite3.dylib). To add this library, make sure that the **General** tab is open. Then, scroll down to the **Linked Frameworks and Libraries** section and click on the + button and add the libsqlite3.dylib library to our project:

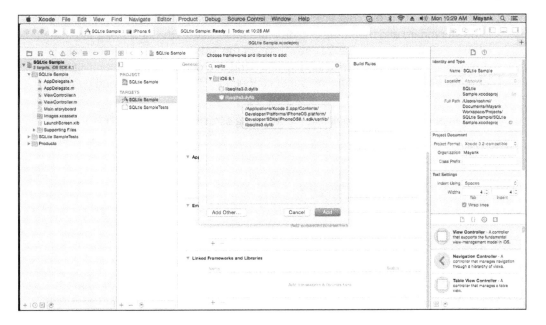

How to do it...

Now, our project is ready to save the SQLite data. For that, we need to write some code. Perform the following steps to update the project as per our requirements:

1. Before we can create a database, we need to import the SQLite framework at the top of the screen. Now, declare a variable for `sqlite3` and create a `NSString` property to store the database path. Within the main Xcode project navigator, select the `DatabaseViewController.h` file and modify it as follows:

   ```
   #import <UIKit/UIKit.h>
   #import <sqlite3.h>

   @interface DatabaseViewController : UIViewController

   @property (strong, nonatomic) NSString *pathDB;
   @property (nonatomic) sqlite3 *sqlDB;
   @end
   ```

2. Now it's time to design our user interface for the SQLite iPhone application. Select the storyboard and then drag and drop the table view to the view along with the table view cell. For `tableviewcell`, select the **Subtitle** style from the attribute and give it an identifier (for example, cell); and add one button on the navigation bar (`style -> add`). Make an outlet connection of the table view and button to `ViewController.h`.

   The final screen should look like the following screenshot:

3. Now we will have to connect the table view outlet with the code. For that, press *Ctrl* and drag the `tableView` object to the `ViewController.h` file. Once connected, you should be able to see the connection in the dialog box with an establish outlet connection named `tableView`. Repeat the steps to establish the action connections for all the other UI components in view.

   Once the connections are established for all, on completion of these steps, the `ViewController.h` file should read as follows:

```
#import <UIKit/UIKit.h>
#import <sqlite3.h>

@interface ViewController: UIViewController
<UITableViewDataSource, UITableViewDelegate>
```

```
@property (strong, nonatomic) NSString *pathDB;
@property (nonatomic) sqlite3 *sqlDB;
@property (weak, nonatomic) IBOutlet UITableView *tableView;

(IBAction)navigateToNextView:(id)sender;

@end
```

4. Now we need to check whether the database file already exists or not. If it is not there, then we need to create the database, path, and table. To accomplish this, we need to write some code in our `viewDidLoad` method. So go to the `ViewController.m` file and modify the `viewDidLoad` method as follows:

```
- (void)viewDidLoad {
  [super viewDidLoad];
  NSString *directory;
  NSArray *dirPaths;
  dirPaths = NSSearchPathForDirectoriesInDomains(
    NSDocumentDirectory, NSUserDomainMask, YES);
  NSArray *paths = NSSearchPathForDirectoriesInDomains(
    NSDocumentDirectory, NSUserDomainMask, YES);
  NSString *documentsDirectory = [paths objectAtIndex:0];
  NSString *dataPath = [documentsDirectory
    stringByAppendingPathComponent:@"/MyFolder"];
  if (![[NSFileManager defaultManager]
    fileExistsAtPath:dataPath])
    [[NSFileManager defaultManager]
      createDirectoryAtPath:dataPath
        withIntermediateDirectories:NO attributes:nil
          error:nil];
  directory = dirPaths[0];
  _pathDB = [[NSString alloc]
  initWithString: [directory
    stringByAppendingPathComponent:@"employee.db"]];
  NSFileManager *filemgr = [NSFileManager
    defaultManager];
  if ([filemgr fileExistsAtPath: _pathDB] == NO)
  {
    const char *dbpath = [_pathDB UTF8String];
    if (sqlite3_open(dbpath, &_sqlDB) == SQLITE_OK)
    {
      char *errMsg;
      const char *sql_stmt =
        "CREATE TABLE IF NOT EXISTS EMPLOYEE (ID INTEGER
        PRIMARY KEY AUTOINCREMENT, NAME TEXT,
        DESIGNATION TEXT)";
```

```
    if (sqlite3_exec(_sqlDB, sql_stmt, NULL, NULL,
      &errMsg) != SQLITE_OK)
    {
      NSLog(@"Failed to create table");
    }
    sqlite3_close(_sqlDB);
  } else {
    NSLog(@"Failed to open/create table");
  }
}
```

The code in the preceding method performs the following tasks:

First, we identify the paths available in our directory and store them in an array (`dirPaths`). Then, we create the instance of `NSFileManager` and use it to detect whether the database has been created or not. If the file has not been created, then we create the database via a call to the SQLite `sqlite3_open()` function and create the table as well. And at last, we close the database.

5. Create a new class named `SecondViewController`. Go to the storyboard and drag one view controller to the canvas. Design the UI according to the following screenshot:

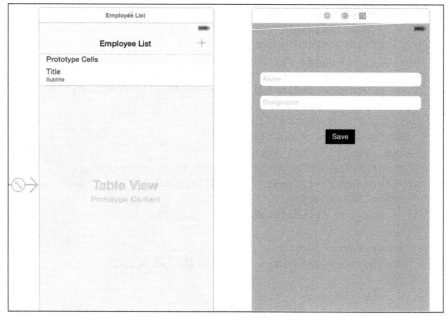

Now make an outlet and action connection for this.

6. Create some properties in `SecondViewController.h`:

```
@property (strong, nonatomic) NSString *pathDB;
@property (nonatomic) sqlite3 *sqlDB;
```

7. In addition, to save the data first, we need to check whether the database is open or not; if it is open, then write a query to insert the data in our table. After writing the data in our table, clear the text present in the text fields. At last, close the database as well.

   In order to implement this behavior, we need to modify the `save` method:

```
- (IBAction)saveButton:(id)sender {
  sqlite3_stmt *statement;
  const char *dbpath = [_pathDB UTF8String];
  if (sqlite3_open(dbpath, &_sqlDB) == SQLITE_OK)
  {
    NSString *sqlQuery = [NSString stringWithFormat: @"INSERT
      INTO EMPLOYEE (name, designation) VALUES
        (\"%@\",\"%@\")",self.nameTextField.text,
          self.designationTextField.text];

    const char *sqlSTMNT = [sqlQuery UTF8String];
    sqlite3_prepare_v2(_sqlDB, sqlSTMNT,
    -1, &statement, NULL);
        if (sqlite3_step(statement) == SQLITE_DONE)
        {
            self.nameTextField.text = @"";
            self.designationTextField.text = @"";
        } else {
            NSLog(@"Failed to add contact");
        }
        sqlite3_finalize(statement);
        sqlite3_close(_sqlDB);
  }
}
```

8. Now we want to populate this saved data in `tableview`. To achieve this task, go to `ViewController.m` and create one mutable array using the following line of code and initialize it in `viewDidLoad`. This array is used to save our SQLite data:

```
NSMutableArray *dataFromSQL;
```

9. In the same class, add the following code to the action button to push the view to new view:

```
SecondViewController *secondView = [self.storyboard
  instantiateViewControllerWithIdentifier:
    @"SecondViewController"];
secondView.pathDB = self.pathDB;
secondView.sqlDB = self.sqlDB;
[self.navigationController pushViewController:secondView
  animated:YES];
```

10. Create the `viewWillAppear` method in the `ViewController` class and modify it as in the following code:

```
- (void)viewWillAppear:(BOOL)animated
{
  [super viewWillAppear:animated];
  [dataFromSQL removeAllObjects];
  [self.tableView reloadData];
  const char *dbpath = [_pathDB UTF8String];
  sqlite3_stmt *statement;

  if (sqlite3_open(dbpath, &_sqlDB) == SQLITE_OK)
  {
    NSString *querySQL = [NSString stringWithFormat:
      @"SELECT name, designation FROM EMPLOYEE"];

    const char *query_stmt = [querySQL UTF8String];

    if (sqlite3_prepare_v2(_sqlDB, query_stmt, -1,
      &statement, NULL) == SQLITE_OK)
    {
      while (sqlite3_step(statement) == SQLITE_ROW)
      {
        NSString *name = [[NSString alloc]
          initWithUTF8String:
        (const char *) sqlite3_column_text(statement, 0)];

        NSString *designation = [[NSString alloc]
          initWithUTF8String:
        (const char *) sqlite3_column_text(statement, 1)];
        NSString *string = [NSString
        stringWithFormat:@"%@,%@", name,designation];
        [dataFromSQL addObject:string];
      }
      [self.tableView reloadData];
```

```
        sqlite3_finalize(statement);
    }
    sqlite3_close(_sqlDB);
  }
}
```

In the preceding code, first we open SQLite and then we fire a query SELECT name, designation FROM EMPLOYEE through which we can get all the values from the database. After that, we will make a loop to store the values one by one in our mutable array (dataFromSQL). After storing all the values, we will reload the tableview.

11. Modify the tableview data source and the delegate method as follows, in ViewController.m:

```
- (NSInteger)tableView:(UITableView *)tableView
  numberOfRowsInSection:(NSInteger)section {
    return dataFromSQL.count;
}


(UITableViewCell *)tableView:(UITableView *)tableView
  cellForRowAtIndexPath:(NSIndexPath *)indexPath
{

  static NSString *CellIdentifier = @"GBInboxCell";
  UITableViewCell *cell = [tableView
    dequeueReusableCellWithIdentifier:CellIdentifier];

  if (cell == nil) {
    cell = [[UITableViewCell alloc]
      initWithStyle:UITableViewCellStyleSubtitle
        reuseIdentifier:CellIdentifier];
  }

  if (dataFromSQL.count>0) {
    NSString *string = [dataFromSQL objectAtIndex:
      indexPath.row];
    NSArray *array = [string componentsSeparatedByString:
      @","];
    cell.textLabel.text = array[0];
    cell.detailTextLabel.text =array[1];


  }
  return cell;
}
```

12. The final step is to build and run the application. Feed the contact details in the second page of the application:

13. Now click on the **Save** button to save your contact details to the database.

14. Now, when you go back to the **Employee List** view, you will see the newly saved data in your table view:

Learning about core data

Saving data is an important feature of almost all iOS apps. Apple has provided one more option for developers to save data using the native database, which is called Core data. This section will introduce us to the basics of core data with an example project. In the previous section, you learned about SQLite, and in this section, we will migrate from the same database to core data. The objectives will be to allow the user to save data in the database and fetch it.

Getting ready

In order to proceed with this exciting topic, we will create a new project. To create a new project, open Xcode and perform the following functions:

1. From the **File** menu, select the option to create a new project.

2. Then, select the **Single View Application** option. Make sure that while creating the project, the **Use Core Data** checkbox is checked.

3. Then, click on **Next** and select a location to save your project. Now you will see that the Xcode has created a new project for us and will launch the project window.

While having a closer look at the project files on the right-hand panel, we will observe an additional file named `CoreData.xcdatamodeld`. This is the file that is going to hold all our entity-level descriptions for data models.

How to do it...

Perform the following steps to learn the implementation of core data in iOS applications:

1. To use the core data in an application, first we will have to create the entity for the `CoreData` application. For this, select the `CoreData.xcdatamodeld` file and load it in the entity editor:

2. In order to add new entities to the data model, click on the **Add Entity** button, which is located in the bottom bar. Once the entity is created, double-click on **New Entity** and change its name to Employee.

3. Once the entity is created, we can now add attributes for it in order to represent the data for the model. Click on the **Add Attribute** button (**+**) to add attributes to the entity. For each attribute that is added, set the appropriate type of data. Similarly, add all the attributes one after another.

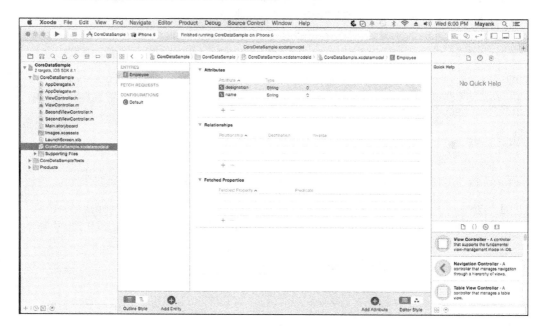

4. Now we will create the user interface for the application and will connect all the UI components with their respective `IBOutlets` and actions. Here, we are following a design similar to our SQLite sample; over and above design, we will add one more class, which is `SecondViewController`:

5. Check all the `IBOutlet` connections in the `SecondViewController.h` file, all the properties should have a dark circle indicator. Then we need to import the `AppDelegate.h` file at the top. After all the changes the class will look similar to following code:

```
#import <UIKit/UIKit.h>
#import "AppDelegate.h"

@interface SecondViewController : UIViewController
@property (weak, nonatomic) IBOutlet UITextField *nameTextField;
@property (weak, nonatomic) IBOutlet UITextField *designationTextField;

- (IBAction)saveButton:(id)sender;
@end
```

6. When the user touches the **Save** button, the `save` method is called. We need to implement the code to obtain the managed object context and create and store managed objects containing the data entered by the user. Select the `ViewController.m` file, scroll down and save the method, and implement the code as follows:

```
- (IBAction) saveButton: (id) sender {
  AppDelegate *appDelegate = [[UIApplication
    sharedApplication] delegate];

  NSManagedObjectContext *context =
  [appDelegate managedObjectContext];
  NSManagedObject *newContact;
  newContact = [NSEntityDescription
    insertNewObjectForEntityForName:@"Employee"
      inManagedObjectContext:context];

    [newContact setValue: _nameTextField.text
      forKey:@"name"];
    [newContact setValue: _designationTextField.text
      forKey:@"designation"];
  _nameTextField.text = @"";
  _designationTextField.text = @"";
}
```

In the preceding code, we created a shared instance of [UIApplication sharedApplication]. This code snippet will always give the same instance every time. Afterwards, it will create nsmanagedobject with the entity we created in CoreData.xcdatamodel, and then it will set the values for the keys.

7. Now we need to populate the same data in our `tableview`. To achieve that, go to `ViewController.m`, create one mutable array (for example, `dataFromCore`) to save our data from `CoreData` to local, and modify `viewWillAppear`:

```
- (void) viewWillAppear: (BOOL) animated
{
  [super viewWillAppear:animated];

  [dataFromCore removeAllObjects];
  [self.tableView reloadData];
  AppDelegate *appDelegate = [[UIApplication
    sharedApplication] delegate];
  NSManagedObject *matches = nil;
```

```objc
NSManagedObjectContext *context = [appDelegate
  managedObjectContext];

NSEntityDescription *entityDesc = [NSEntityDescription
  entityForName:@"Employee"
  inManagedObjectContext:context];

NSFetchRequest *request = [[NSFetchRequest alloc] init];
[request setEntity:entityDesc];

NSError *error1 = nil;
NSArray *results = [context executeFetchRequest:request
  error:&error1];
if (error1 != nil) {
  NSLog(@"%@", error1);
}
else {
  for (matches in results) {
    NSLog(@"%@.....%@", [matches valueForKey:@"name"],
      [matches valueForKey:@"designation"]);
    NSString *string = [NSString stringWithFormat:
      @"%@,%@", [matches valueForKey:@"name"],
        [matches valueForKey:@"designation"]];
    [dataFromCore addObject:string];

  }
}
[self.tableView reloadData];
}
```

Again, it will create the shared instance of [UIApplication sharedApplication] and it will return the same instance as previously. Now, we are fetching the data from NSFetchRequest using NSEntityDescription and saving the result in one array. After that, we retrieve the values from the keys, one by one, and add them to our mutable array (dataFromCore), and at last, reload our tableview.

8. Our tableview methods should look similar to the following code. If they do not, then modify the code:

```objc
- (NSInteger)tableView:(UITableView *)tableView
  numberOfRowsInSection:(NSInteger)section {
  return dataFromCore.count;
}
```

```objc
- (UITableViewCell *)tableView:(UITableView *)tableView
  cellForRowAtIndexPath:(NSIndexPath *)indexPath
{
    static NSString *CellIdentifier = @"GBInboxCell";
    UITableViewCell *cell = [tableView
      dequeueReusableCellWithIdentifier:CellIdentifier];
    if (cell == nil) {
        cell = [[UITableViewCell alloc] initWithStyle:
          UITableViewCellStyleSubtitle reuseIdentifier:
            CellIdentifier];
    }
    if (dataFromCore.count>0) {
        NSString *string = [dataFromCore objectAtIndex:
          indexPath.row];
        NSArray *array = [string
          componentsSeparatedByString:@","];
        cell.textLabel.text = array[0];
        cell.detailTextLabel.text =array[1];

    }
    return cell;
}
```

9. The final step is to build and run the application. Enter the text in the second view and hit **Save** to save your contact details in our database:

Now, when you go back to the **Employee List** view, you will see the newly saved data in your table view as shown in the following screenshot:

2
Social Media Integration

Social media has captured a huge market place in the technological world. This market shift redirects the focus of app developers to social media integration in their applications. In this chapter, our prime focus will be to integrate various social media platforms in our applications. This chapter will further broaden our approach to understand this for analytical usage as well.

In this chapter, we will cover:

- ▸ Integration with Facebook or Twitter
- ▸ Integration with LinkedIn
- ▸ Integration with Instagram

Integration with Facebook

Integration of social sites in an app is very common nowadays. With the release of iOS 6, Apple has introduced a new framework for social media integration known as `Social.framework`. This social framework allows developers to instantly integrate social networking services into applications such as Facebook and Twitter.

Getting ready

We will create an app to demonstrate social media integration. We will start by creating a new iOS app project.

1. Open Xcode and go to **File | New | File** and select **Single View Application** by navigating to **iOS | Application**. In the popup, provide the product name **Social Media Integration.** It should look similar to the following screenshot:

2. Click on **Next** and save the project. After creating the project, we need to add a `Social.framework`. To add this library, go to the `target` action of the project and scroll down to **Linked Frameworks and Libraries** and click on the **Add** button and add `Social.framework` in our project.

How to do it...

Now our project is ready to start. For that, we will first design our storyboard and then we check out some smart codes to integrate with social media. Follow these steps to update the project as per our requirements:

1. Go to the storyboard and design the UI. Add two buttons, one for Facebook integration and another for Twitter integration. Then, link it to our class. Our storyboard looks like the following screenshot:

2. Now, import the `Social.h` class in our `ViewController.h` file:

    ```
    #import<Social/Social.h>
    ```

3. Now, add the following code in `ViewController.m` in the button actions for the Facebook and Twitter buttons, respectively.

    ```
    - (IBAction)facebookButton:(id)sender {
        SLComposeViewController *controller = [SLComposeViewController
    composeViewControllerForServiceType: SLServiceTypeFacebook];
        [controller setInitialText:@"Hurray!! Posting from my app"];
        [self presentViewController:controller animated:YES
                        completion:Nil];
    }

    - (IBAction)twitterButton:(id)sender {
        SLComposeViewController *tweetSheet = [SLComposeViewController
    composeViewControllerForServiceType:
            SLServiceTypeTwitter];
    ```

```
        [tweetSheet setInitialText:@" Share on Twitter -_- "];
        [self presentViewController:tweetSheet animated:YES
    completion:nil];
    }
```

4. Run the application and tap on the Facebook or Twitter button.

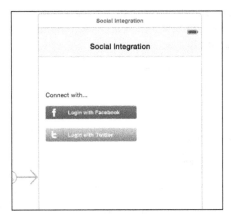

5. Now, you will be able to post on Facebook or Twitter from your app.

It will work as shown in the preceding screenshot on Twitter as well. Try it.

Integrating with LinkedIn

Now, we'll use `Social.framework` to integrate LinkedIn into our application.

Getting ready

For LinkedIn integration, we will continue with the preceding project. Open the project and redesign the storyboard. Add another button with the Facebook and Twitter buttons. Drag one more view controller from the inspector element and drop it on the canvas. Design the new view according to the following screenshot. There, we have two labels, two buttons, and one text view.

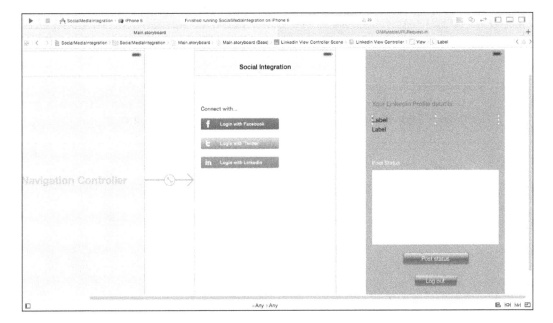

How to do it...

1. Add a new class for the new view (for example, `LinkedinViewController`). Make a connection for the UI element, and our `LinkedinViewController.h` file looks like the following screenshot:

```objectivec
#import <UIKit/UIKit.h>
@interface LinkedinViewController : UIViewController

@property (strong, nonatomic) IBOutlet UILabel *nameLabel;
@property (strong, nonatomic) IBOutlet UILabel *designatioLabel;
@property (strong, nonatomic) IBOutlet UITextView *statusTextfield;

- (IBAction)postButton:(id)sender;
- (IBAction)logoutButton:(id)sender;

@end
```

2. Now, we need to register our app on the LinkedIn developer site. Go to `https://www.linkedin.com/secure/developer` and add your application. There we will get an API key and secret key, which will need in our code.

3. After registering our app, we need to download the third-party library for integration of LinkedIn. Go to `https://s3.amazonaws.com/oodles-site-files/LoginApp.zip` and download the `LinkedIn` library.

4. After downloading the library, add the library folder and its classes to our project.

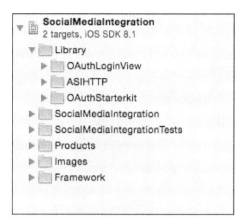

5. Go to `ViewController.m`, import `LinkedinViewController.h`, and add the following code in the following `LinkedIn` action button:

```
- (IBAction)likedinButton:(id)sender {
    LinkedinViewController *linkedinView = [self.storyboard instan
tiateViewControllerWithIdentifier:@"LinkedinViewController"];
    [self.navigationController pushViewController:linkedinView
animated:YES];


}
```

The preceding code will push the application to the next view.

6. Now, import the following classes in `LinkedinViewController.h` and create two methods for the profile and the network API call. We also need an object for the `OauthLoginView` class:

```
#import <UIKit/UIKit.h>
#import "OAuthLoginView.h"
#import "JSONKit.h"
#import "OAConsumer.h"
#import "OAMutableURLRequest.h"
#import "OADataFetcher.h"
#import "OATokenManager.h"
#import <Foundation/NSNotificationQueue.h>

@property (nonatomic, strong) OAuthLoginView *oAuthLoginView;

- (void)profileApiCall;
- (void)networkApiCall;
```

7. Our final `.h` file looks like the following screenshot:

```objc
#import <UIKit/UIKit.h>
#import "OAuthLoginView.h"
#import "JSONKit.h"
#import "OAConsumer.h"
#import "OAMutableURLRequest.h"
#import "OADataFetcher.h"
#import "OATokenManager.h"
#import <Foundation/NSNotificationQueue.h>

@interface LinkedinViewController : UIViewController

@property (strong, nonatomic) IBOutlet UILabel *nameLabel;
@property (strong, nonatomic) IBOutlet UILabel *designatioLabel;
@property (strong, nonatomic) IBOutlet UITextView *statusTextfield;
@property (nonatomic, strong) OAuthLoginView *oAuthLoginView;

- (IBAction)postButton:(id)sender;
- (IBAction)logoutButton:(id)sender;

- (void)profileApiCall;
- (void)networkApiCall;


@end
```

8. Add some frameworks that are needed to use this library:

```
SystemConfiguration.framework
CFNetwork.framework
libz.dylib
```

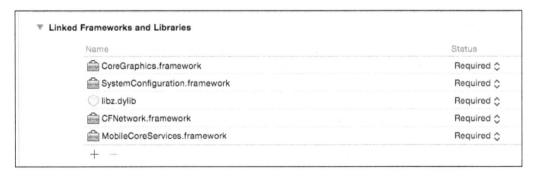

9. Now, go to `LinkedinViewController.m` and modify the `viewDidLoad` method:

```objc
- (void)viewDidLoad {
[super viewDidLoad];

    self.oAuthLoginView = [[OAuthLoginView alloc]
        initWithNibName:nil bundle:nil];
```

```
    [[NSNotificationCenter defaultCenter] addObserver:self
  selector:@selector(loginViewDidFinish:)
  name:@"loginViewDidFinish"
  object:self.oAuthLoginView];

[self presentViewController:self.oAuthLoginView
  animated:YES completion:nil];
}
```

In the preceding code, we created one notification, which calls `loginViewDidFinish` when login is completed. This will present `oAuthLoginView`.

10. Now, add the remaining code for the `LinkedinViewController.m` file:

```
-(IBAction) postButton: (id) sender {
  [self.statusTextfield resignFirstResponder];
  NSURL * url = [NSURL URLWithString:
    @ "http://api.linkedin.com/v1/people/~/shares"];
  OAMutableURLRequest * request = [
    [OAMutableURLRequest alloc] initWithURL: url
    consumer: self.oAuthLoginView.consumer
    token: self.oAuthLoginView.accessToken
    callback: nil
    signatureProvider: nil
  ];
  NSDictionary * update = [
    [NSDictionary alloc] initWithObjectsAndKeys: [
      [NSDictionary alloc]
      initWithObjectsAndKeys:
      @ "anyone", @ "code", nil
    ], @ "visibility",
    self.statusTextfield.text, @ "comment", nil
  ];
  [request setValue: @ "application/json"
    forHTTPHeaderField: @ "Content-Type"
  ];
  NSString * updateString = [update JSONString];
  [request setHTTPBodyWithString: updateString];
  [request setHTTPMethod: @ "POST"];
  OADataFetcher * fetcher = [ [OADataFetcher alloc] init];
  [fetcher fetchDataWithRequest: request
```

```
        delegate: self
        didFinishSelector: @selector(postUpdateApiCallResult:
          didFinish:)
        didFailSelector: @selector(postUpdateApiCallResult:
          didFail:)
    ];
}
- (IBAction) logoutButton: (id) sender {
    NSString * tokenKey = @ "";
    [[NSUserDefaults standardUserDefaults] setObject:
        tokenKey forKey: @ "TokenKey"
    ];

    [self.navigationController popViewControllerAnimated:
      YES];
    UIAlertView * logoutMessageAlert = [
      [UIAlertView alloc] initWithTitle: @ "Alert"
      message: @ "Logout Successfully !"
      delegate: self cancelButtonTitle: nil otherButtonTitles:
        @ "OK", nil
    ];
    [logoutMessageAlert show];
}

- (void) loginViewDidFinish: (NSNotification * )
      notification {
    [[NSNotificationCenter defaultCenter] removeObserver:
        self
    ];

    // We're going to do these calls serially
    // just for easy code reading.
    // They can be done asynchronously
    // Get the profile, then the network updates
    [self profileApiCall];
}
- (void) profileApiCall {
    NSURL * url = [NSURL URLWithString:
      @ "http://api.linkedin.com/v1/people/~"];
    OAMutableURLRequest * request = [
      [OAMutableURLRequest alloc] initWithURL: url
      consumer: self.oAuthLoginView.consumer
      token: self.oAuthLoginView.accessToken
      callback: nil
```

```objc
      signatureProvider: nil
  ];
  [request setValue: @ "json"
    forHTTPHeaderField: @ "x-li-format"
  ];
  OADataFetcher * fetcher = [
    [OADataFetcher alloc] init
  ];
  [fetcher fetchDataWithRequest: request
    delegate: self
    didFinishSelector: @selector(profileApiCallResult:
      didFinish: )
    didFailSelector: @selector(profileApiCallResult:
      didFail: )
  ];
}
- (void) profileApiCallResult: (OAServiceTicket * )
    ticket didFinish: (NSData * ) data {
  NSString * responseBody = [
    [NSString alloc] initWithData: data
    encoding: NSUTF8StringEncoding
  ];
  NSDictionary * profile = [responseBody
    objectFromJSONString];
  if (profile) {
    self.nameLabel.text = [
      [NSString alloc] initWithFormat: @ "%@ %@",
        [profile objectForKey: @ "firstName"],
      [profile objectForKey: @ "lastName"]
    ];
    self.designatioLabel.text = [profile objectForKey: @
      "headline"];
  }
  // The next thing we want to do is call the network updates
  [self networkApiCall];
}
- (void) profileApiCallResult: (OAServiceTicket * )
    ticket didFail: (NSData * ) error {
  NSLog(@ "%@", [error description]);
```

```
}
- (void) networkApiCall {
  NSURL * url = [NSURL URLWithString: @
    "http://api.linkedin.com/v1/people/~/network/updates?
      scope=self&count=1&type=STAT"];
  OAMutableURLRequest * request = [
    [OAMutableURLRequest alloc] initWithURL: url
    consumer: self.oAuthLoginView.consumer
    token: self.oAuthLoginView.accessToken
    callback: nil
    signatureProvider: nil
  ];
  [request setValue: @ "json"
    forHTTPHeaderField: @ "x-li-format"
  ];
  OADataFetcher * fetcher = [
    [OADataFetcher alloc] init
  ];
  [fetcher fetchDataWithRequest: request
    delegate: self
    didFinishSelector: @selector(networkApiCallResult:
      didFinish: )
    didFailSelector: @selector(networkApiCallResult:
      didFail: )
  ];
}
- (void) networkApiCallResult: (OAServiceTicket * )
    ticket didFinish: (NSData * ) data {
  NSString * responseBody = [
    [NSString alloc] initWithData: data
    encoding: NSUTF8StringEncoding
  ];
  NSDictionary * person = [[[[responseBody objectFromJSONString]
    objectForKey: @ "values"]
      objectAtIndex: 0]
        objectForKey: @ "updateContent"]
          objectForKey: @ "person"];
  if ([person objectForKey: @ "currentStatus"])
  {
  }
  else {
```

```
    }
    [self dismissViewControllerAnimated: YES completion: nil];
}
- (void) networkApiCallResult: (OAServiceTicket * ) ticket
  didFail: (NSData * ) error {
    NSLog(@ "%@", [error description]);
}
- (void) postUpdateApiCallResult: (OAServiceTicket * ) ticket
  didFinish: (NSData * ) data {
    // The next thing we want to do is call the network updates
    [self networkApiCall];
}
- (void) postUpdateApiCallResult: (OAServiceTicket * ) ticket
  didFail: (NSData * ) error {
    NSLog(@ "%@", [error description]);
}
```

11. Now, go to `OAuthLoginView.m` and add the secret and API keys in the `initLinkedInAPi` method:

```
- (void)initLinkedInApi
{
    apikey = @"API_KEY";
    secretkey = @"API_SECRET";

    self.consumer = [[OAConsumer alloc] initWithKey:apikey
                                        secret:secretkey
                                        realm:@"http://api.linkedin.com/"];

    requestTokenURLString = @"https://api.linkedin.com/uas/oauth/
        requestToken";
```

12. The most important thing is that we need to make our project non-ARC because the library that we are using is written in non-ARC. So, go to **Build Settings** and search for `Automatic Reference Counting` and turn off ARC.

▼ Apple LLVM 6.0 - Language - Objective C	
Setting	⚛️ SocialMediaIntegration
Objective-C Automatic Reference Counting	No ↕

13. Now, our application is ready to integrate with LinkedIn. Run the application and hit the LinkedIn integration button.

14. Now, it will navigate us to the LinkedIn login page. Enter the login ID and password.

15. Hurray! You are logged in. Now you can post on your LinkedIn wall through your app.

16. After posting, you can easily logout by hitting the **Logout** button.

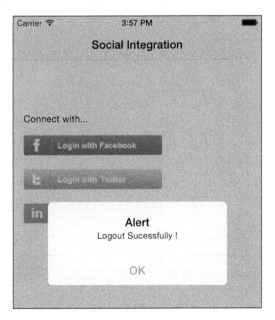

Integration with Instagram

Now, we'll use `Social.framework` to integrate Instagram into our application.

Getting ready

For Instagram integration, we will continue with the preceding project. Open the project and redesign the storyboard. Add another button after Facebook, Twitter, and LinkedIn. Drag one more view controller from the inspector element and drop it onto the canvas. Design the new view according to the following screenshot. Drag two buttons from the interface builder and name them Login and Logout respectively.

How to do it...

1. Add a new file class in our project and name it `InstagramLoginViewController`. Make this class the owner of the newly created view controller. Then, link the button to this class as the action method

2. Now, we have to register our app to the Instagram Developers site as we did in previous integrations. Go to `http://instagram.com/developer/register/` and register your app.

3. After a successful registration, Instagram will give you a client ID; save it somewhere, as we will need it in our code.

4. Go to `https://github.com/crino/instagram-ios-sdk/blob/master/README.md` and download the Instagram iOS SDK.

5. After downloading, copy the Instagram iOS SDK into our project, as shown here:

6. Go to our `.pch` project file and import `IGConnect.h` so it will automatically import in our whole project, so we do not need to define it in every class.

7. Now, modify the `AppDelegate` class in `AppDelegate.h` and the `define` property of `AppDelegate.h` of the `Instagram` class.

   ```
   @property (strong, nonatomic) Instagram *instagram;
   ```

8. In `AppDelegate.m`, define our client ID and write some code as follows:

   ```
   #define APP_ID @"whatever your App ID define here"
   ```

9. In the `didFinishLaunching` method, initialize the following Instagram property:

   ```
   self.instagram = [[Instagram alloc] initWithClientId:APP_ID
   delegate:nil];
   ```

10. Add two methods as follows:

    ```
    - (BOOL)application:(UIApplication *)application
      handleOpenURL:(NSURL *)url {
      return [self.instagram handleOpenURL:url];
    }

    - (BOOL)application:(UIApplication *)application
      openURL:(NSURL *)url sourceApplication:(NSString *)
        sourceApplication annotation:(id)annotation {
        return [self.instagram handleOpenURL:url];
    }
    ```

11. Now, go to `InstagramLoginViewController.h` and import `AppDelegate.h` and conform one protocol, `IGSessionDelegate`.

12. Then, go to `InstagramLoginViewController.m` and modify the `viewDidLoad` method as follows:

    ```
    self.logoutbutton.enabled = NO;
    ```

13. In the `login` action method, write the following code, which will authorize Instagram with Safari:

```
AppDelegate* appDelegate = (AppDelegate*)[UIApplication
  sharedApplication].delegate;
  [appDelegate.instagram authorize:[NSArray
    arrayWithObjects:@"comments", @"likes", nil]];
  self.logoutbutton.enabled = YES;
  self.loginbutton.enabled = NO;
```

14. After all this, write this method in the same class:

```
    #pragma - IGSessionDelegate

  -(void)igDidLogin {
    NSLog(@"Instagram did login");
    // here I can store accessToken
    AppDelegate* appDelegate = (AppDelegate*)[UIApplication
sharedApplication].delegate;
    [[NSUserDefaults standardUserDefaults] setObject:appDelegate.
instagram.accessToken forKey:@"accessToken"];
    [[NSUserDefaults standardUserDefaults] synchronize];

    IGListViewController* viewController =
[[IGListViewController alloc] init];
    [self.navigationController    pushViewController:viewControll
er animated:YES];
    }

  -(void)igDidNotLogin:(BOOL)cancelled {
    NSLog(@"Instagram did not login");
    NSString* message = nil;
    if (cancelled) {
        message = @"Access cancelled!";
    } else {
        message = @"Access denied!";
    }
    UIAlertView* alertView = [[UIAlertView alloc]
initWithTitle:@"Error" message:message delegate:nil
cancelButtonTitle:@"Ok" otherButtonTitles:nil];

    [alertView show];
    }

  -(void)igDidLogout {
    NSLog(@"Instagram did logout");
    // remove the accessToken
    [[NSUserDefaults standardUserDefaults] setObject:nil
forKey:@"accessToken"];
    [[NSUserDefaults standardUserDefaults] synchronize];
    }
```

```
-(void)igSessionInvalidated {
 NSLog(@"Instagram session was invalidated");
 }
```

These are the delegate methods of the `IGSessionDelegate` protocol. As the name specifying their functionality, in the `igDidLogin` and `igDidLogout` methods, we are saving an access token in `NSUserDefault` so that we can check whether the user is already logged in or not by using the `isSessionValid` method of the `IGSession` class.

15. Add the following code in the `logout` action method:

```
AppDelegate* appDelegate = (AppDelegate*)[UIApplication
sharedApplication].delegate;
[appDelegate.instagram logout];

UIAlertView *alert = [[UIAlertView alloc]
initWithTitle:@"Successfully Logout" message:nil delegate:self
cancelButtonTitle:@"OK" otherButtonTitles:nil, nil];
[alert show];
self.loginbutton.enabled = YES;
self.logoutbutton.enabled = NO;
```

16. Compile and run the code.

17. Tap on the Instagram login button, it will redirect you to the login and logout view of `InstagramViewController`. Tap on the login button, it will redirect you to Safari to log in to Instagram. Enter your credentials and hurray! Now you are logged in with Instagram as well.

After login, your **Logout** button will be enabled, so you can logout by clicking on that button.

3

Integrating Data Analytics

In this chapter, we will cover the following topics:

- ▶ Google Analytics
- ▶ Flurry Analytics
- ▶ Flurry with data mining
- ▶ Integrating Mixpanel

Introduction

When any application for mobile is developed, the performance of the application becomes its priority. The developer and the owner of the application need to keep track of the usage of application worldwide.

So the question is, how do we track the user from the world, off the shelf? To solve this problem, Google and Yahoo provide frameworks named Google Analytics and Flurry, respectively. So, in this chapter, we will cover these topics and we will see how to track our app and see how many users are using our app along with their locations.

Google Analytics

Google Analytics provides a service to developers that generates detailed statistics about the usage of the application in real time. Google Analytics can be used to track mobile apps as well as websites.

Getting ready

To develop a mini app using Google Analytics, start by creating a new project. Open Xcode and go to **File** | **New** | **File**, then navigate to **iOS** | **Application** | **Single View Application**. Within the popup, provide the product name `AnalyticsSample`.

How to do it...

1. Before implementing Google Analytics, please make sure that you have a Google account. Go to `http://www.google.co.in/analytics/` and sign in with your credentials.

2. After login, you will see the home screen. At the top, you will see the admin button near the home button. Go to the admin page. There, you will see three columns. The first column is for accounts. It will currently be empty. Click on **Accounts** and you will see a **Create new account** option; click on it, as shown in the following screenshot:

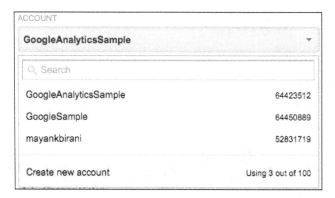

3. On this screen, you will see the two options: **Website** and **Mobile app**. Select **Mobile app**, and it will ask you the account name; you can give any name (giving the same name as the app name is recommended). Fill in the credentials and click on the **Get Tracking ID** button.

4. Here, you will get your tracking ID as shown in the following screenshot. Save it anywhere and then download the Google Analytics iOS SDK.

5. Now, move to the Xcode and add some files to your project from the SDK package:

    ❑   GAI.h

    ❑   GAITracker.h

    ❑   GAITrackedViewController.h

    ❑   GAIDictionaryBuilder.h

    ❑   GAIFields.h

    ❑   GAILogger.h

    ❑   libGoogleAnalyticsServices.a

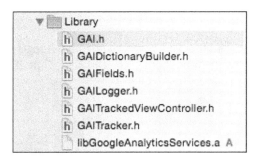

6. Add the following frameworks into your project:

    ❑   AdSupport.framework

    ❑   CoreData.framework

    ❑   SystemConfiguration.framework

    ❑   libz.dylib

    ❑   libsqlite3.dylib

7. To initialize a tracker, go to `AppDelegate.m` and import `GAI.h`, and in the `didFinishLaunching` method, write the following code before returning to `Yes`:

```
[GAI sharedInstance].trackUncaughtExceptions = YES;
[GAI sharedInstance].dispatchInterval = 20;
[[[GAI sharedInstance] logger]
  setLogLevel:kGAILogLevelVerbose];


// Initialize tracker.id<GAITracker> tracker = [[GAI
  sharedInstance] trackerWithTrackingId:@"UA-64450889-1"];
```

8. If the user wants to track which view or screen of the app is currently active, this facility is also provided by the Analytics SDK. Just follow the code to accomplish screen tracking. Go to `ViewController.m` and integrate `GAITrackedViewController.h,` and in `viewDidLoad`, write this line of code:

```
self.screenName = @"Home Screen";
```

9. Now, run the project. You will see the following output in your console:

```
GET: https://ssl.google-analytics.com/collect/av=1.0&cid=1a419b2d-5abd-45a1-
a8a3-7cc35f5cac46&tid=UA-64450889-1&a=2021670972&dm=x86_64&cd=Home
+Screen&t=screenview&aid=com.exilant.GoogleAnalyticsSample&ul=en&_u=.etno&ds=app&sr=750x1334&v=1&_s=2
&_crc=0&an=GoogleAnalyticsSample&_v=mi3.1.2&ht=1435138770217&qt=19979&z=4534304260846321384
2015-06-24 15:09:52.065 GoogleAnalyticsSample[4730:116385] INFO: GoogleAnalytics 3.12 -
[GAIBatchingDispatcher didSendHits:response:data:error:] (GAIBatchingDispatcher.m:226): Hit(s)
dispatched: HTTP status -1
2015-06-24 15:09:52.065 GoogleAnalyticsSample[4730:116467] INFO: GoogleAnalytics 3.12 -
[GAIBatchingDispatcher deleteHits:] (GAIBatchingDispatcher.m:529): hit(s) Successfully deleted
2015-06-24 15:09:52.075 GoogleAnalyticsSample[4730:116467] INFO: GoogleAnalytics 3.12 -
[GAIBatchingDispatcher didSendHits:] (GAIBatchingDispatcher.m:237): 1 hit(s) sent
```

10. After you see **1 hit(s) sent** in your console, go to the Google Analytics web page, and then go to **Reporting** (near the home button) | **Real Time** (left panel) | **Overview.**

11. There, you will see the number of active users and other statistics as well.

So, in this way, we can track the number of users, the date of joining, and the location along with many logs.

Flurry Analytics

Like Google Analytics, Flurry is used to track the user's session and number of active users, along with the location and full statistics, but Flurry also provides another good functionality, ads. Through Flurry, we can show ads in our app. The three types of ads that Flurry provides are Native, Interstitial, and Banner ads.

Let's start the session with Flurry Analytics.

Getting ready

To develop a mini app using Flurry Analytics, start by creating a new project. Open Xcode and go to **File** | **New** | **File** and then to **IOS** | **Application** | **Single View Application**. In the popup, provide the product name `FlurryIntegration`.

How to do it...

1. Go to the Yahoo Developer site (`https://dev.flurry.com/secure/login.do`) and sign in with your credentials. If you are new, then create an account.

2. Next, it will ask you the platform on which our application is built upon. Select the **iPhone** option.

3. Give the application's name and in **Categories**, select **Education**. Then, click on **Create app**.

4. Now, they will give you a unique application key. Save it somewhere and then download the Flurry SDK, which is just below the unique key:

5. After downloading the SDK, open the SDK and add the `Flurry` and `FlurryAds` files to our project as shown in the following screenshot:

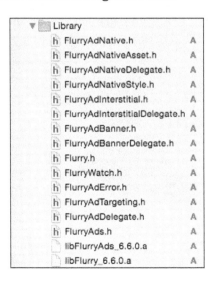

6. Now, go to `AppDelegate.h` and import the `Flurry.h` file:

   ```
   #import "Flurry.h"
   ```

7. Then, move to `AppDelegate.m` and start a session with the `didFinishLaunching` method:

   ```
   [Flurry startSession:@"R6NQXQ2WGP78MPBW9VBV"];
   ```

```
- (BOOL)application:(UIApplication *)application didFinishLaunchingWithOptions:(NSDictionary *)launchOptions {
    [Flurry startSession:@"R6NQXQ2WGP78MPBW9VBV"];
    return YES;
}
```

8. Run the code. Whoa! We get an error. We need to import one more framework to start the session. Go to the project target | **General** | **Linked Frameworks and Libraries** and add `SystemConfiguration.framework`.

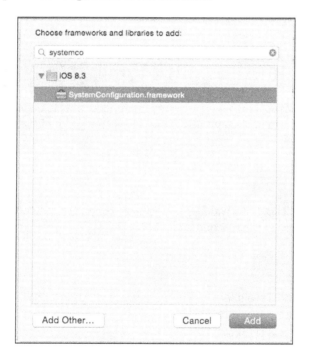

9. Again, run the project. This time, we can successfully build the code. In the console, it will show that a session has started.

```
2015-06-30 12:08:14.932 FlurryIntegration[13949:346376] Flurry: Starting session on Agent Version
[Flurry_iOS_154_6.6.0]
```

10. After a successful build, go to our Flurry developer web page, where we got the unique key. Go to the home page from the top.

Here, we can see the number of active users, as shown in the following screenshot. Click on the application; it will redirect you to the statistics page where you can see every detail of your application. Explore it.

Unique Users	Active Users	▼ Sessions
1	1	1
1	1	4
0	0	0
0	0	0

Flurry with data mining

As we previously discussed, Flurry also provides the functionality to display ads. In this section, we will see how to implement this in our app with a good, efficient approach. Let's dive into full screen ads.

Getting ready

For this session, we will continue with the project that we made in the preceding section. Open the `FlurryIntegration` Xcode project.

How to do it...

1. Go to the Yahoo Developer site (`https://dev.flurry.com/secure/login.do`) and sign in with your credentials.

2. Here, you will see the home page. Near the home button, you can see the **Publishers** button; click on it. From the name itself, it is obvious that it is made to publish something.

3. This will navigate you to a different page. On this page, you will see four panels to the left. Open the third panel, **Inventory**, and select **Ad Spaces**.

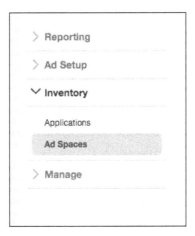

4. Here, perform basic setup for the ad. Name the ad, choose an application, and under **Placement**, select **Fullscreen**, and then click on **Save Ad Space**.

5. Now, go to Xcode. Select **Storyboard**, and drag one button to the view canvas and link an action method to the `ViewController` class.

6. Select `ViewController.m` and import two classes, `FlurryAdInterstitial.h` and `FlurryAdInterstitialDelegate.h`, and write some code as given here:

```
@property FlurryAdInterstitial* adInterstitial;
```

7. Now modify `viewDidLoad`:

```
- (void)viewDidLoad {
  [super viewDidLoad];

  self.adInterstitial = [[FlurryAdInterstitial alloc]
    initWithSpace:@"FlurryAdd"] ;
  self.adInterstitial.adDelegate = self;
  [self.adInterstitial fetchAd];
}
```

8. In the `action` method, present the ad using the following code:

```
- (IBAction)showAD:(id)sender {

  if ([self.adInterstitial ready] == YES) {
    [self.adInterstitial presentWithViewController:self];
  } else {
    // if the ad is not ready, fetch the ad
    self.adInterstitial = [[FlurryAdInterstitial alloc]
      initWithSpace:@"FlurryAdd"];
    self.adInterstitial.adDelegate = self;
    [self.adInterstitial fetchAd];
  }
}
```

We use the ad space name that we gave at the time of creation. It will detect our ad from this ad space name.

9. It does not end here. For ad support, there is a framework by Apple called `AdSupport.framework`; add it along with some more frameworks:

   ❑ `MediaPlayer.framework`

   ❑ `libz.dylib`

10. Now, it's time to run the code. Press the button, and see the magic.

Integrating Mixpanel

Mixpanel provides a wide range of services to developers. It is impossible to get a realistic view of our app without mobile analytics. The number of users of the application is an important metric because it shows the health of our application. Mixpanel is a good platform to explore analytics for the mobile platform. Tracking an event from your application is an easy task in Mixpanel. Let's start the session with Mixpanel.

Getting ready

To develop a mini app using Google Analytics, start by creating a new project. Open Xcode and go to **File** | **New** | **File** and select **iOS** | **Application** | **Single View Application**. In the popup, provide the product name IntegrateMixPanel.

How to do it...

1. Mixpanel integration can be done with CocoaPods and without CocoaPods. Most of the time doing it with CocoaPods is recommended because it simplifies version updates and dependency management. But there are various reasons when a user can't use CocoaPods sometimes it is because of security checks and sometimes users don't have admin privileges. Here, we will see how to integrate Mixpanel without CocoaPods so that it's better and more accessible for every developer.

2. Go to the terminal and Git; clone the latest version of the Mixpanel SDK for iPhone. Write the following command to successfully clone from Git:

```
git clone https://github.com/mixpanel/mixpanel-iphone.git
```

3. Your terminal will start the cloning and you can see the process. Your terminal will look like the following screenshot:

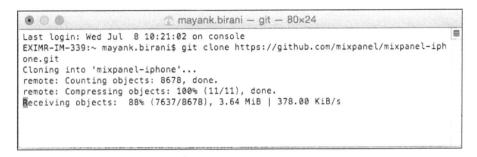

4. After the successful cloning, open the SDK and you will see the `mixpanel` folder; copy all the files from the folder to our project.

5. Now, add some frameworks that are necessary because they need to support the Mixpanel SDK.

    ❏   `Libicucore.dylib`

    ❏   `QuartzCore.framework`

    ❏   `CoreGraphics.framework`

    ❏   `Accelerate.framework`

    ❏   `CoreTelephony.framework`

    ❏   `SystemConfiguration.framework`

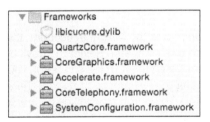

6. Now, we need a Mixpanel account for the application token and to see our application's activity. Go to the official Mixpanel site (`https://mixpanel.com/`) and create a new account if you don't have one; otherwise, log in.

7. Once you are logged in, you can create a new project for your application. Give the name of the project and click on **Create new project**.

8. After creating the project, at the bottom of the page, you will see three buttons as shown in the following screenshot. Select the first one; you will get a token ID for the application.

9. Move to our Xcode project, import `Mixpanel.h` into `AppDelegate.m`, and add the initializer `Mixpanel` to the `didFinishLaunching` method:

    ```
    #import "Mixpanel.h"
        - (BOOL)application:(UIApplication *)application
          didFinishLaunchingWithOptions:(NSDictionary
            *)launchOptions {

        [Mixpanel sharedInstanceWithToken:MIXPANEL_TOKEN];
        }
    ```

    Give your Mixpanel token as a string in place of `MIXPANEL_TOKEN`.

    ```
    - (BOOL)application:(UIApplication *)application didFinishLaunchingWithOptions:(NSDictionary *)launchOptions {
        // Override point for customization after application launch.
        [Mixpanel sharedInstanceWithToken:@"4ec0103091b08a1216ed0df1fb92902b"];
        return YES;
    }
    ```

10. Now, run the project and go to the Mixpanel account. It will ask whether you are using an iPhone device or a simulator. Select the device you are using; let's assume it's an iPhone simulator. To connect to the simulator, you need to press the **Option** button along with holding the left mouse button on the simulator screen for 5 seconds. After doing this, your simulator will be connected to Mixpanel as shown in the following screenshot:

11. We can see our statistics on the left panel. Go and explore, there are several options.

12. Now, we will see how to track events, assuming we have a button click event in our sample. Go to **Storyboard** and drag a button into the view canvas and make an action connection.

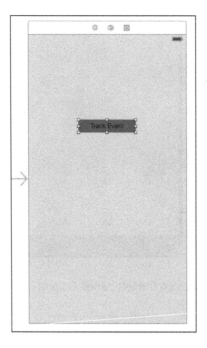

13. Next, run the application again. Go to the Mixpanel account and you will also see the **Track Event** button there as a virtual view. Right-click on the button and set the name of the event, then click on the **Deploy** button on the right.

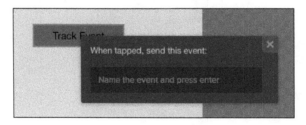

14. Run the application again. Click on the **Track Event** button in the simulator and you will see the event action in the Mixpanel account, as in the following screenshot:

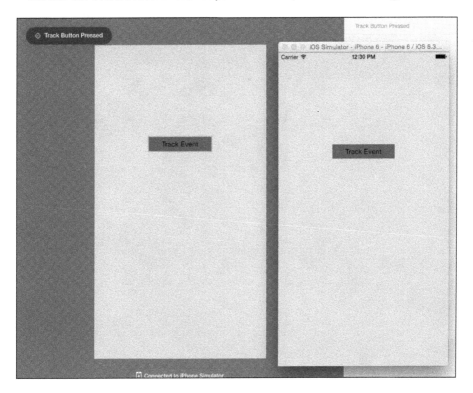

15. If you look in the **Segmentation** option on the left panel, you can track the user and the events. As an example, our track button event will be tracked there and you can see the date of the specific events.

4

App Distribution and Crash Reporting

In this chapter, we will be focusing on the following topics:

- ▸ Setting up and integrating TestFlight
- ▸ Integrating HockeyApp
- ▸ Hockey app for crash reporting

In *Chapter 3*, *Integrating Data Analytics*, you learned about various analytics tools and their features. In this chapter, our primary focus will be on the application distribution life cycle and its crash management system. Currently, there are two major iOS distribution platforms, which are TestFlight and HockeyApp.

Setting up and integrating TestFlight

TestFlight was a major app distribution channel for all iOS developers and has been recently acquired by Apple. After the acquisition, it underwent various changes, for example now you can't access and upload build files from their website. Apple has deeply integrated the beta testing of the app through iTunes Connect. In this section, you will learn about how to deeply integrate and distribute our beta builds for testing using TestFlight.

Getting ready

In order to learn how to distribute an app via TestFlight, our primary requirement would be to create a sample app that we can distribute to beta testers. Please perform these steps as the prerequisite for uploading the beta build to the TestFlight app:

1. Before deep diving, we have to set up a project. Let's create a simple iOS application; open Xcode, navigate to **File | New | Project**, and select **Single View Application**, then click on **Next**.

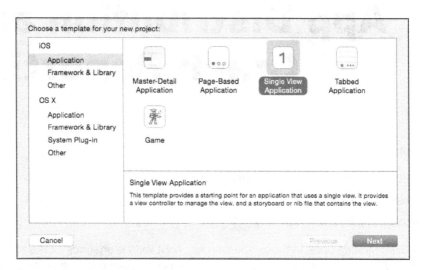

2. Now just type in the name of the application and bundle ID; then, click on **Next**.

How to do it...

1. To use TestFlight, you have to create an application on iTunes Connect. Now, let's first create the app ID on the Apple developer portal in order to access the same bundle ID on iTunes Connect. Navigate to `https://developer.apple.com/devcenter/ios/index.action`. Once logged in with your Apple ID and password, create a new bundle ID. Provide all the required details as shown in the following screenshot:

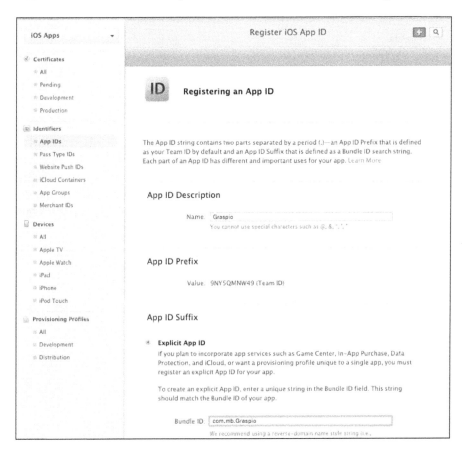

Apple's integration with TestFlight has several advantages; now, for the beta release, you will not have to create separate ad hoc provisioning profiles. Also, you will not have to provide any device details; now, TestFlight builds will work on all the devices that are invited.

2. Now, sign in to your iTunes Connect account and create a new application. Fill in all the required details and make sure you select the correct bundle ID from the drop-down menu and click on **Create,** as shown in the following screenshot:

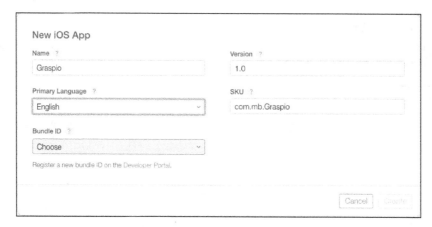

3. Once the app is successfully created, you will be redirected to the main dashboard page of the app. You will not need to provide any further information on the iTunes Connect account. Now, click on the **Prerelease** tab. This will be the place where all the builds will be listed.

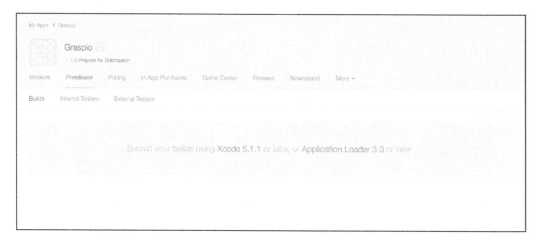

4. Now, we are all set for the build upload. Let's come back to our Xcode and release a build. Click on the **Build Settings** tab for your app target, and make sure that **Provisioning Profile** is set to **Automatic**. The **Code Signing Identity** parameter for the release should be iOS Developer for **Any iOS SDK**, which is the default setting when you create a project, as shown in the following screenshot:

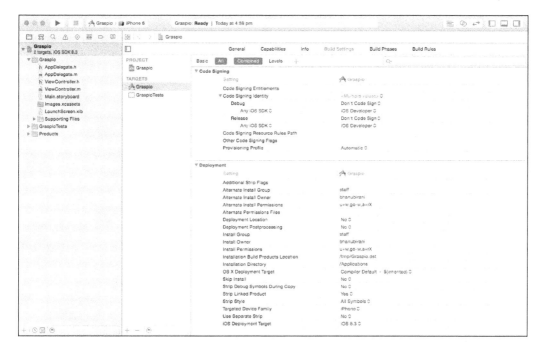

5. Now, make sure you have added all the required app icons for the app. This is mandatory for any app before the beta release. If we don't provide these icons, then Xcode is going to fail with an error.

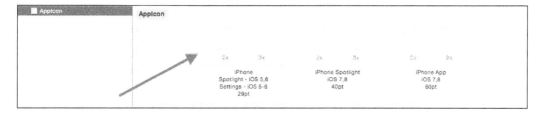

6. Once we have the app icons in place, we can proceed and archive the build for the beta release. To create an archive with release configuration, navigate to **Product | Archive**. Make sure that you selected the iOS device from the device settings. You are not allowed to create an archive with iOS simulators. The settings should be as shown in the following screenshot:

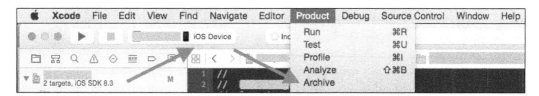

7. Once the archive is successfully finished, you will be automatically taken to the **Organizer** window. This window should show the list of archives you have generated so far. On the right-hand side, you should see the **Submit to App Store** button, as shown in the following screenshot:

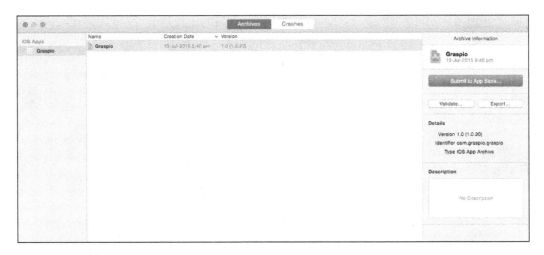

8. Now, once you click on **Submit to App Store**, Xcode will show you the warning page, which should look something similar to the following screenshot. Xcode is unable to get your provisioning profile information and as a result, it throws this warning. To resolve this warning, you can click on the **Try Again** button.

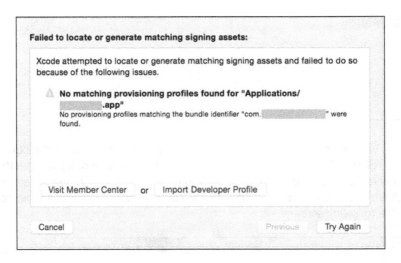

9. Once you click on **Try Again**, Xcode will autogenerate the ad hoc certificate and provisioning profile based on the iTunes Connect data. The name of the autogenerated provisioning profile should start with XC, which symbolizes that it has been autogenerated by Xcode. Now, once you have successfully completed this step, you should be able to see the progress bar for uploading your binary to iTunes Connect, as shown in the following screenshot:

10. You will receive a confirmation popup with a success message once the build is successfully uploaded to iTunes Connect as shown in the following screenshot:

11. Now, we will navigate back to our browser and check the build availability in the iTunes Connect portal. Click on the **Prerelease (Builds)** version and you should be able to see something similar to the following screenshot:

12. Now, click on the **TestFlight** tab. Fill in all the required metadata and **What to Test**; after that the page should look similar to the following screenshot:

13. After saving the metadata information, you can go back to the previous page and enable the **TestFlight Beta Testing** switch, which is on the right side of the page.

14. Now, you can add the multiple internal and external testers. To add external testers, you can click on the **External Tester** tab and it should look similar to the following screenshot:

Currently, we can see there are two external testers already added; you can add up to 1000 testers for each build.

15. To add new external testers, click on **+** in the **External Tester** tab. This should display a popup with two options, **Add New Tester** and **Add Existing Tester**. This should look similar to the following screenshot:

16. Now, click on **Add New Tester** and fill in all the required information. The form should look similar to the following screenshot:

17. Once you save any tester data, the tester will receive an invitation e-mail to test your app.

18. Now, your tester can download the **TestFlight** app from the App Store on their iOS device. After this, they will simply have to click on the invitation e-mail and install the app on their device.

19. In this section, you successfully learned how to distribute our app using TestFlight, and we are certainly sure that everyone will find it more easy and streamlined to distribute our apps using TestFlight.

Integrating HockeyApp

HockeyApp is a widely used app distribution platform. It allows users to deploy and distribute apps. HockeyApp provides you with a number of features. In this section, you will learn the necessary steps to get started with HockeyApp's integration with iOS apps.

Getting ready

These are the preliminary requirements to get started with HockeyApp:

1. Go to `http://hockeyapp.net/features/`.

2. We need to sign up on the HockeyApp portal.

3. We need to activate the account and have to fill in all the preliminary information in the account details section. The final account details page should look similar to the following screenshot:

This completes our registration on HockeyApp.

1. Once you sign up, you will be using the general account, using which you can add any apps. You can start using the one-month trial. This trial will allow you to use all the features for one month.

2. Go to the account settings using the top-right button of the web page as shown in the following screenshot:

3. Now, navigate to **Upgrade** and click on **Start Trial**

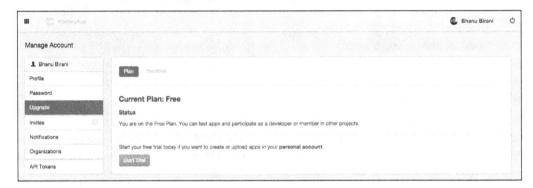

4. Now, you can navigate back to the dashboard by clicking on the box-shaped icon on the top left of the screen.

5. You should now be able to see the following screen:

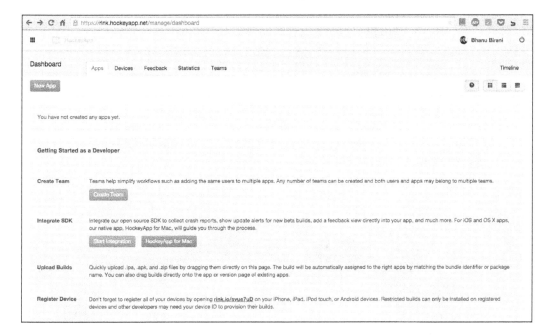

This completes our registration process.

How to do it...

In this section, you will learn about the process of integrating the HockeyApp SDK with our app. First, we will start by creating the app on HockeyApp. Please follow the steps to create the app:

1. The first way to create the app is as follows:

   1. You can directly drag and drop your build, that is, your `.ipa` file, onto the dashboard. The drag and drop will start the uploading process, and you can see the progress bar once the build is dropped. This should look similar to the following screenshot:

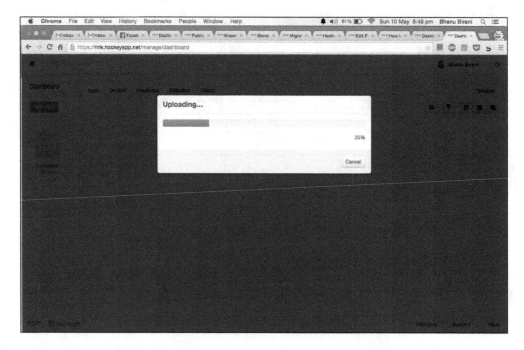

   2. The build will be automatically assigned to the right apps by matching the bundle identifier or package name.

3. Once the upload is finished, your status will change to **Processing**. Once the processing is completed, you will be redirected to the version page. This should look similar to the following screenshot:

2. You can also create the app using the following process:

1. Once you click on the **New App** button, you should be able to see the following pop-up window:

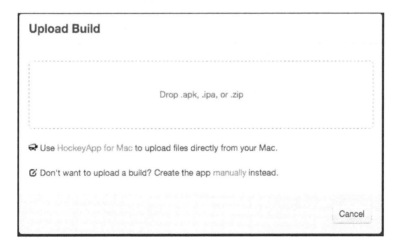

2. Now, instead of dropping the build onto the pop-up window, click on **manually**. This will redirect you to the page where you will have to fill in all your app-related details. The page should look similar to the following screenshot:

3. Now, fill in the details and click on the **Save** button; you will be redirected to the app page, wherein you can check all your app-related details. The page should look similar to the following screenshot:

3. So far, we have seen how to create an app on the HockeyApp dashboard. Now, we will manually integrate the HockeyApp SDK in our app in order to start collecting crash reports and distributing the app. For this, we need to download the HockeyApp SDK from `http://hockeyapp.net/releases/`.

4. Unzip the package once the file is downloaded. It should look similar to the next screenshot.

5. Now, we can copy the folder to our project directory on the disk.

6. To add it to the Xcode project, drag and drop `HockeySDK.embeddedframework` into your framework section. This should look similar to the following screenshot:

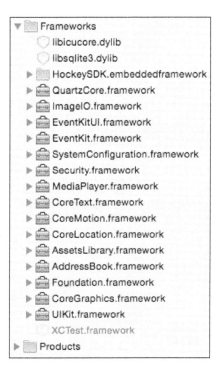

7. Now, we have to drop the other files in our Xcode project section too. For this, make sure that the **Create groups** option is selected for any added folders and set the check mark for your target. Then, click on **Finish**, as shown in the following screenshot:

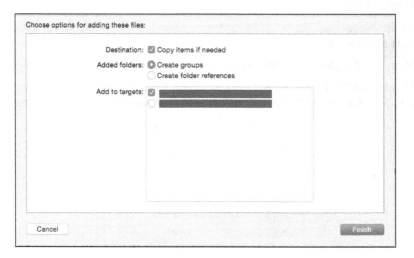

8. Now, go to your Project Navigator.

9. Select your app target, navigate to **Build Phases**, and expand **Link Binary with Libraries,** as shown in the following screenshot:

10. Now, you will have to add the system frameworks that are missing. Here is the laundry list of the frameworks needed:

   - ❑ `AssetsLibrary`
   - ❑ `CoreText`
   - ❑ `CoreGraphics`
   - ❑ `Foundation`
   - ❑ `MobileCoreServices`
   - ❑ `QuartzCore`
   - ❑ `QuickLook`
   - ❑ `Security`
   - ❑ `SystemConfiguration`
   - ❑ `UIKit`

11. Now, browse to `HockeySDK.embeddedframework/HockeySDK.framework/Versions/A/Resources/HockeySDKResources.bundle` and drag and drop the `HockeySDKResources.bundle` into your Xcode project.

12. With all the preceding steps, we have now completed the process of moving all the assets and frameworks to the project. Now, we can add some code to initiate HockeyApp. To do so, we have to open our `AppDelegate.m` file.

13. Add the following line of code right after all the import statements:

    ```
    #import <HockeySDK/HockeySDK.h>
    ```

14. Now, in the `application:didFinishLaunchingWithOptions:` method, add the following lines of code:

    ```
    [[BITHockeyManager sharedHockeyManager]
    configureWithIdentifier:@"APP_ID"];
    // Configure the SDK in here only!
    [[BITHockeyManager sharedHockeyManager] startManager];
    [[BITHockeyManager sharedHockeyManager].authenticator
    authenticateInstallation];
    ```

15. You will also need `APP_ID` to initiate the HockeyApp logging. HockeyApp will assign you a unique hash such as `APP_ID`, which can be accessed from the app detail page, as shown in the following screenshot:

16. Now, we are all set to release our first version to HockeyApp. You will need to set up the code signing for the project. Make sure that the provisioning profile holds all the device IDs to which we want to send the build.

17. To upload the build to HockeyApp, we will install the Mac GUI client for HockeyApp. We will download the dmg from `http://hockeyapp.net/apps/`.

18. Install the app once the download is complete.

19. Now, open the app and log in with your username and password. It should look similar to the following screenshot after login:

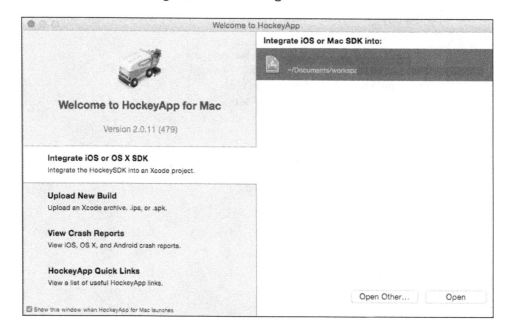

20. Now, go to Xcode and navigate to **Product | Archive** to archive the build for distribution.

21. Since we have installed our desktop app of HockeyApp, we will receive a popup on the Mac with the **Upload** button. Make sure you have the HockeyApp app open while archiving. The popup should look similar to the following screenshot:

22. Click on **Upload** and you should be able to see something like the following screenshot:

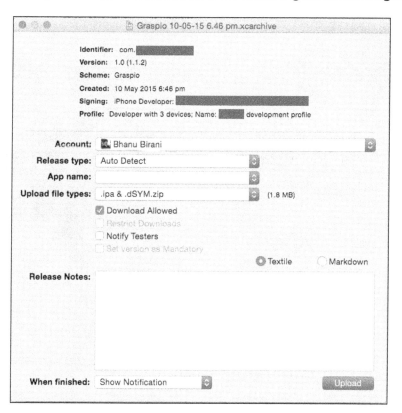

You can write the **Release notes** while uploading the build, which will be visible to users on the download page of the app.

23. The build that we have uploaded is bundled with a developer provisioning profile. This will only allow the registered user to be able to download the build. To check the list of provisioned devices, go to the dashboard and click on the app we created. You should be able to see the page listing all the uploaded versions of the app. The page should look similar to the following screenshot:

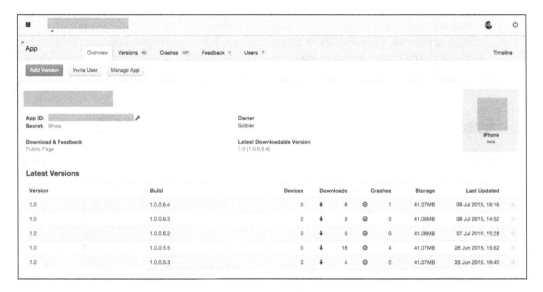

24. Once you click on any version from the version list, you will be redirected to the **Details** page. The **Details** page should have a section with the header **Provisioning**, which should show the number of devices provisioned in the profile. The page should look similar to the following screenshot:

25. Once you click on the device list, you will be redirected to a page with all the list of devices that are provisioned. The device-listing table should have a column that displays the **UDID** of the device and a column labeled **Owner** that should display the owner name of the device. On the first sign-up, you should not be able to see any owners in the list. You need to send sign-up invites to all the team members. Once the team has signed up successfully, the device name used to sign up will be saved as their name. The final screen should look like the following screenshot:

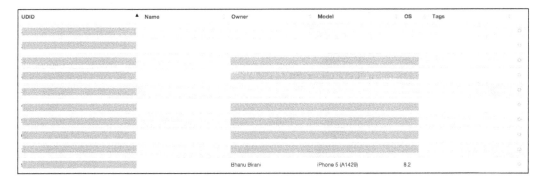

26. To send invites to your team and other beta testers, click on the app and select the **Users** tab from the available options. The page should look like the following screenshot:

27. Click on **Invite User** to send a new invite to the user. The invite page should look like the following screenshot:

28. A notification e-mail is sent whenever a new user is uploaded for all the users to know who has signed up and has been successfully added to the team. All the users who are registered as `Developer` and `Member` will also receive e-mail notifications for crashes and feedback.

29. Finally, we have successfully integrated HockeyApp in our app for distribution and reporting.

HockeyApp for crash reporting

In this section, we will dive in deep with the crash reporting system of HockeyApp. HockeyApp has a very robust crash reporting system; we will take a step-by-step look at integrating it with our app.

Getting ready

It is a prerequisite to read and understand the previous section. We will be starting from where we left off in the previous section.

How to do it...

Perform the following steps in order to learn about the integration of crash reporting system in our apps:

1. We already integrated the HockeyApp SDK in our app in the previous section. HockeyApp provides us with the `BITCrashManager` class to support an efficient crash reporting system.

2. By default, crash reporting is enabled in HockeyApp, which will send the crashes the next time the app is launched.

3. If you need to disable HockeyApp crash reporting, you can add the following code to `applicationDidFinishLoading`:

```
[[BITHockeyManager sharedHockeyManager]
  configureWithIdentifier:@"APP_IDENTIFIER"];
[[BITHockeyManager sharedHockeyManager]
  setDisableCrashManager: YES];
[[BITHockeyManager sharedHockeyManager] startManager];
```

4. By default, the user will be prompted to send a crash report once the application is launched after the crash. The user can decline to send a crash report as well. However, if you need to auto-send a crash report without any user interaction, you can use the following line of code:

```
[[BITHockeyManager sharedHockeyManager]
configureWithIdentifier:@"APP_IDENTIFIER"];
```

```
[[BITHockeyManager sharedHockeyManager].crashManager
    setCrashManagerStatus: BITCrashManagerStatusAutoSend];
[[BITHockeyManager sharedHockeyManager] startManager];
```

5. In the preceding code snippet, we changed the status of the crash manager to `BITCrashManagerStatusAutoSend`. Once this flag is set, crash reports are asynchronously sent automatically on the next launch.

6. You can also send various other components with the crash reports to specifically study the user's crash data. To enable such features, you need to subscribe to `BITHockeyManagerDelegate`.

7. `BITHockeyManagerDelegate` uses the following methods of delegate to report additional data:

```
UserID: - (NSString
*)userIDForHockeyManager:(BITHockeyManager
*)hockeyManager componentManager:(BITHockeyBaseManager
*)componentManager;


UserName: - (NSString
*)userNameForHockeyManager:(BITHockeyManager
*)hockeyManager componentManager:(BITHockeyBaseManager
*)componentManager;


UserEmail: - (NSString *)userEmailForHockeyManager:
    (BITHockeyManager *)hockeyManager
      componentManager:(BITHockeyBaseManager
        *)componentManager;
```

Using these methods, you can send the user ID, username, and user e-mail along with crash reports.

8. So far, we have seen some code for integrating crash analytics in our app. Now, we will take a deeper look at the crashes which are reported. Now, click on the app from **Dashboard**. Once the app is opened, you should be able to see the **Overview**, **Versions**, **Crashes**, **Feedback**, and **Users** tabs. It should look like the next screenshot.

9. As per the screenshot, we can see that the app has received 285 crashes and now we would like to read about the crashes of our app. To see the list of crashes, click on the **Crashes** tab. It should look like the following screenshot:

10. If we read the list of crashes, we can clearly observe that this list is sorted by the number of times the crash has appeared in the app. The list also gives us the **Description** header, which tells us the primary exception that caused the crash.

11. Along with crash reporting, HockeyApp also provides basic analytics. You can check the **Downloads** count for each version of the app you have uploaded and the time spent by the user with each version. The analytics should look like the following screenshot:

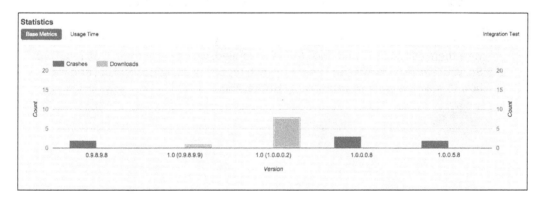

12. Now, in order to check the complete details of the crash along with stack trace, we can click on the crash. This should look like the following screenshot:

13. Now, this is the advantage of uploading the complete distribution archive on HockeyApp. Your crash desymbolication can happen automatically on HockeyApp. As per the preceding screenshot, we can see the class and line number which caused the crash. Also, the complete stack trace is readable and has been decoded by HockeyApp. Using the crash report, a developer can easily check the crash in the `scrollViewDidScroll` method and apply a fix for it.

14. HockeyApp also provides various other features where we can download the logs and view them as RAW file, and once the fix is applied, we can also mark it as **resolved**. So, you learned how we can integrate crash reporting using HockeyApp, and we saw how to read the crashes captured by HockeyApp.

5
Demystifying Crash Reports

In this chapter, we will be focusing on the following topics:

- ▶ Crashlytics integration
- ▶ Desymbolication of crash logs
- ▶ Analyzing crash reports

Crashlytics integration

Along with making applications, developers now also track the crashes and issues after publishing the app. There are a lot of plugins for tracking crashes. Crashlytics is one of them. Crashlytics tracks the crashes and e-mails the ID that is registered to the app.

In this section, we will explore Crashlytics and see how we can add this to our project and how it tracks our crashes.

Getting ready

We will create an app to demonstrate the Crashlytics. We will start by creating a new iOS app project:

1. Open Xcode and go to **File** | **New** | **File**.
2. Navigate to **iOS** | **Application** | **Single View Application**.

3. In the popup, provide the product name `SampleCrashlytics`. It should look like the following screenshot:

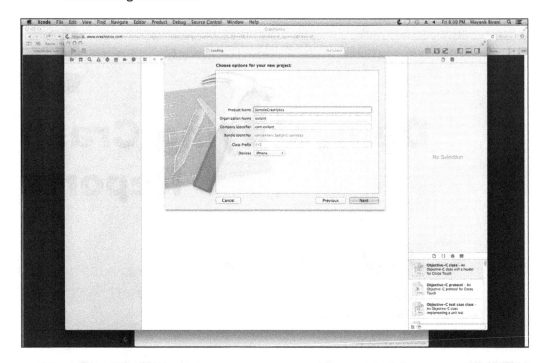

4. Click on **Next**, and save the project.

How to do it...

Now, our project is ready to start. But first, we will design our storyboard, and then we will check out some smart codes to integrate with social media. Perform the following steps to update the project as per our requirements:

1. First, we need to register our app to Crashlytics. Go to `www.crashlytics.com` and register yourself; then, it will ask for the platform for which we want Crashlytics. Select **Xcode** in that option and continue.

2. After registering, we need to download the Crashlytics plugin on our system. Download it from the same page as per the following screenshot:

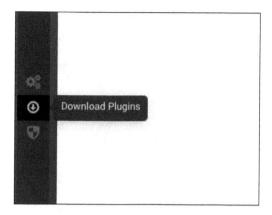

3. After downloading the plugin, we can see the Crashlytics icon on the top of our desktop screen. Clicking on that will show you the following screen:

4. Now, add the Xcode file in the app, and click on **Next**.

5. You can see your app within this app. Click on **Next** to continue.

6. Now, we need to add a run script in our project. To add a run script, go to **Project Settings | Build Phase**. There, you we see the **+** button above the four options. Click on it and add a new run script build phase. Copy the following script from the Crashlytics plugin and add in the new script:

7. After adding the run script, we need to add the Crashlytics framework, which will be autogenerated by the plugin. We just need to drag and drop the framework into our app.

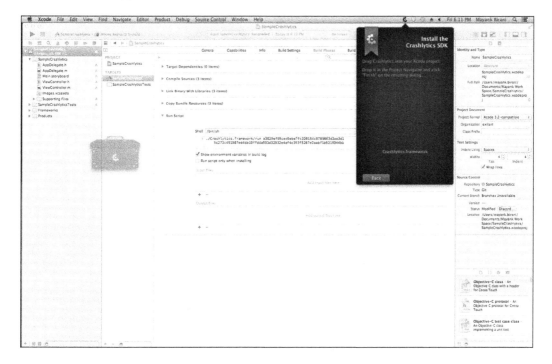

8. Now, go to AppDelegate.m and modify the code.

9. Import the framework, and inside `didFinishLaunch`, write the `Crashlytics` API key, which is generated by the plugin.

10. Now, we have successfully added the Crashlytics app to our project.

11. Now, we need to test whether it's working or not. To do that, we will make our app crash using the following code, which we'll write in `viewDidLoad`:

```
NSMutableArray *arrayForCrash = [[NSMutableArray alloc] init];
NSString *string = [arrayForCrash objectAtIndex:2];
```

12. Now, once you run the app, it will crash, and after some time, you will get an e-mail regarding the crash and your Mac plugin will also show the issues.

Desymbolication of crash logs

Once we have deployed our app, we won't be able to use the Xcode debugger tool to debug it. Our only aim should be to improve the customer's experience with your application. This involves fixing the application crash as soon as possible. This can be achieved by analyzing crash logs to debug the problems.

Receiving crash logs directly from a device without using Xcode

Every build of an app can have multiple crash reports and each crash report contains multiple crash logs, Apple's crash reporting service collects all the crash logs and groups similar crash logs into separate crash reports. The crash logs are collected by Apple only if the app user agrees to share crash data with the app developers. The Apple crash report service collects crash logs from apps running on user devices.

Getting ready

Users can get crash reports from the device itself and send them through e-mail.
To see how to do this, look at the steps in the following section.

How to do it...

1. Open the **Settings** page on the device.
2. Go to **Privacy**, then **Diagnostics & Usage**.
3. Tap on **Diagnostics & Usage Data**:

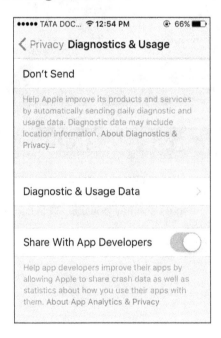

4. Here, you can see the crash logs of all your applications. The logs will be named in the format: AppName_DateTime_DeviceName.

5. Select the desired log. Select the text under **Crash Log via Text** and copy it.

6. Paste the copied text to Mail and send it to the desired e-mail address.

7. Otherwise, we can change the setting to **Automatically Send**, which is used to send the crash log automatically when the device addresses crash in any application.

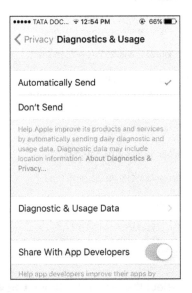

Receiving crash logs from a device by using Xcode

Xcode also allows you to download and see the crashes that happened on your phone. To check the crashes, you can connect your device to your system and open the **Devices** window from Xcode. It also helps you to symbolicate crashes of your app using the dysm file on your machine. Also, this is the only reason Xcode persists all your .xarchive files, which hold the .dysm file. Perform the following steps to analyze the details of the crash logging using Xcode.

How to do it...

1. Open Xcode and plug in the device.

2. Select **Window | Devices** from the menu bar. Now select your device from the left side panel, as shown in the following screenshot:

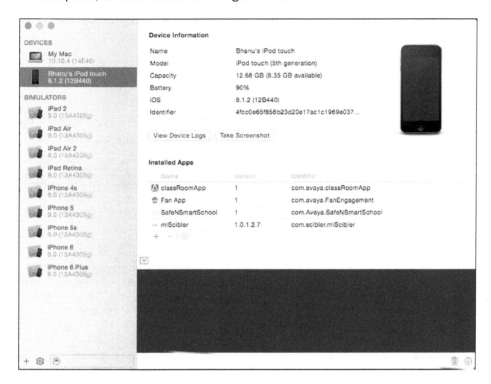

3. Click on the dropdown button at the bottom-left corner of the window. This should open the device console on the screen.

4. Now, in order to save the logs to the file, click on the **Download** button located at the bottom right of the window.

5. In order to check the device logs, click on the **View Device Logs** button located in the **Device Information** section. Now, find the app for which we need to get the crash logs and click on **See the Content**.

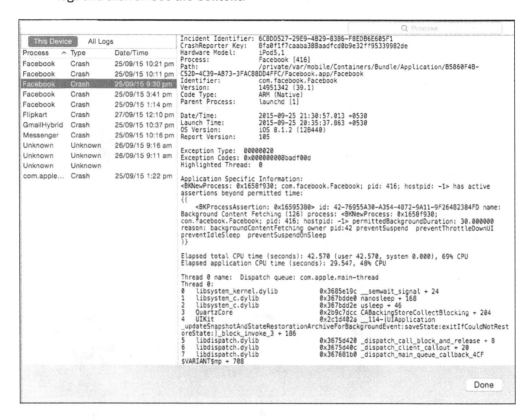

6. Now, to save the crash logs, right-click on the list to the left and select the **Export Log** option, as shown in following screenshot:

Analyzing crash reports

In the previous topics, we have seen various ways to generate and gather crash reports. In this section, our primary focus will be on the ways to analyze the crash reports. This will help us better understand the reasons for crashes and their possible fixes. By default, crashes are not generated in a human-readable format; to read crashes, we need to symbolicate them using the .dysm file.

Getting ready

Open Xcode and navigate to the **Organizer** window. The Organizer window should list all the crashes for all the apps developed by your team. The following steps will help you in analyzing them.

How to do it...

1. Go to **Window | Organizer**. Choose **Crashes**.

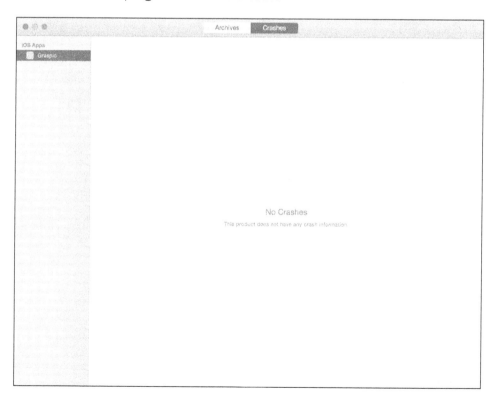

2. Select an app from the left column. The list of applications is fetched from iTunes Connect and it will obtain all the information about every version.

3. In the second column, you can see the versions and the build of your app that encountered a crash.

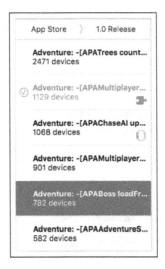

4. To see where exactly in the code there was a crash, navigate to the right panel where you can see **Open in Project...**. Select the log you want to see, and click on **Open in Project...**.

5. It will navigate you to the code where this crash occurred.

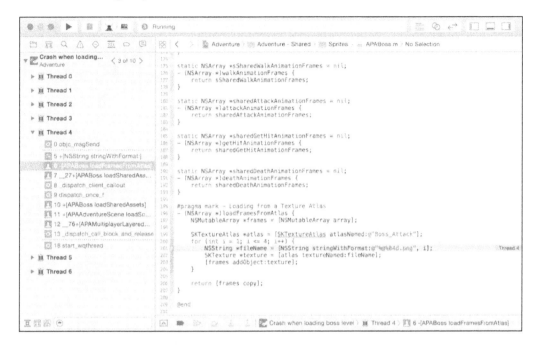

6. Crash Organizer also gives us the facility to mark crash logs as resolved; you can hide resolved crashes. It also allows, user to see the statistics of the crash log for the last two weeks.

6
Forensics Recovery

Forensics recovery is important, as in a few investigation cases there is a need to decrypt information from iOS devices. These devices are in an encrypted form usually. In this chapter, we will focus on various tools and scripts that can be used to read data from the devices under investigation. We are going to cover the following topics:

- ▸ DFU and Recovery modes
- ▸ Extracting and reading data
- ▸ Recovering backups
- ▸ Extracting data from iTunes backups
- ▸ Encrypting and decrypting tools

DFU and Recovery modes

In this section we'll cover both the DFU mode and the Recovery mode separately.

DFU mode

In this section, we will see how to launch the DFU mode, but before that we will explore what DFU means. DFU stands for Device Firmware Upgrade, which means this mode is used specifically with iOS upgrades. This is a mode where a device can be connected with iTunes and still do not load the iBoot boot loader. Your device screen will be completely black in DFU mode because neither the boot loader nor the operating system is loaded. DFU bypasses the iBoot so that you can downgrade your device.

How to do it...

We need to follow these steps in order to launch a device in DFU mode:

1. Turn off your device.

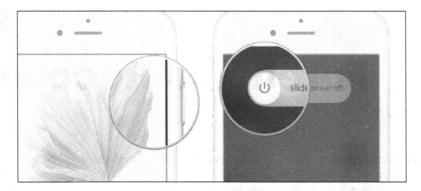

2. Connect your device to the computer.

3. Press your Home button and the Power button, together, for 10 seconds.

4. Now, release the Power button and keep holding the Home button till your computer detects the device that is connected.

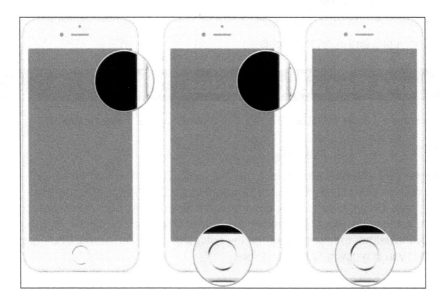

5. After some time, iTunes should detect your device.

6. Make sure that your phone does not show any Restore logo on the device; if it does, then you are in Recovery mode, not in DFU.

7. Once your DFU operations are done, you can hold the Power and Home buttons till you see the Apple logo in order to return to the normal functioning device. This is the easiest way to recover a device from a faulty backup file.

Recovery mode

In this section, you will learn about the Recovery mode of iOS devices. To dive deep into the Recovery mode, we first need to understand a few basics such as which boot loader is used by iOS devices, how the boot takes place, and so on. We will explore all such concepts in order to simplify our understanding of the Recovery mode. All iOS devices use the iBoot boot loader in order to load the operating system. The iBoot's state, which is used for recovery and restore purposes, is called Recovery mode. iOS cannot be downgraded in this state as iBoot is loaded. iBoot also prevents any other custom firmware from flashing into a device unless it is a jailbreak, that is, "pwned".

How to do it...

The following are the detailed steps to launch the Recovery mode on any iOS device:

1. You need to turn off your iOS device in order to launch the Recovery mode.

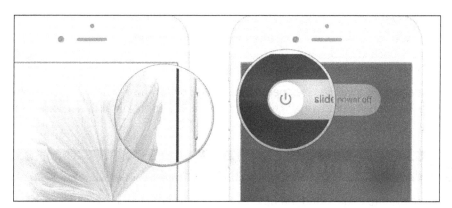

2. Disconnect all the cables from the device and remove it from the dock if it is connected.

3. Now, while holding the Home button, connect your iOS device to the computer using the cable.

4. Hold the Home button till you see the Connect to iTunes screen.

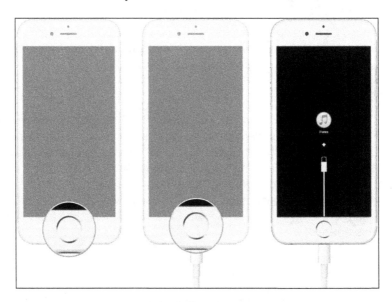

5. Once you see the screen, you have entered the Recovery mode.

6. Now you will receive a popup in your Mac saying "iTunes has detected *your iDevice* in recovery mode".

7. Now you can use iTunes to restore the device in the Recovery mode. Make sure your data is backed up because the recovery will restore the device to Factory Settings. You can later restore from the backup as well.

8. Once your Recovery mode operations are complete, you will need to escape from the Recovery mode. To escape, just press the Power button and the Home button concurrently for 10-12 seconds.

Extracting and reading data

It is very important to understand the basics of the file system before you learn about the ways of extracting and reading data from the iOS device disk. All Apple operating systems use the same file system, hierarchical file system (HFS).

This file system works with the 512 byte-formatted block scheme. To categorize it further, these blocks are divided into two parts: allocation blocks and logical blocks. Logical blocks are available on the volume numbered from the first to the last block. These blocks remain static on the disk. Allocated blocks work with a different strategy; they can be grouped together to utilize the HFS more efficiently. The file structure includes the **Allocation** and **Attributes Files**, along with the **Volume Header** and **Catalog Files**, and so on.

Getting ready

To understand the extraction well, we will study a little about some of the headers of the HFS file system in detail.

The HFS+ Volume Header

For the HFS format disk, the boot blocks are sectors 0 and 1 of the volume. All the information of the HFS volume structure is stored in volume headers. After partition, these volume headers are stored in the boot blocks, which are reserved for boot. The size of this header is 1,024 bytes. The backup of the volume header is stored in the last 1,024 bytes of the volume, to overcome any failure of the operating system. This backup can be used for many purposes and the user can also use it to repair a disk if the original header is corrupted by any circumstances.

A wide variety of volume data is stored in these volume headers such as the size of allocation blocks and time stamp blocks, which are primarily used to know when the volume was created and modified. It also contains the directory path of the volume structure files.

The HFS+ Allocation File

The Allocation File is used to retrieve the free and used space of the disk in the HFS format. This file allows us to know whether the block is free or is being used. All this information is stored in a bitmap with a bit, which indicates the status of the block; this bit is called the clear bit. If the bitmap for the block is zero, then the block is free, else it's used.

The HFS+ Extents Overflow File

The Extents Overflow File is used to track all the allocation blocks for a file. This file is used to save a list of all the allocation blocks in the proper order for every file. The balanced tree format is used to save this information.

The HFS+ Catalog File

In order to get the folder and its hierarchy information for each volume, we use the HFS+ catalog file. This file consolidates and saves all the information and metadata for each volume. This metadata has information about modified timestamps, created timestamps, and also includes information about the access control of the file. This file also uses the balanced tree format to allocate files. In this format, nodes are used to refer files and folders. This file also helps to retain the order of the map, the header, the leaf nodes, and the index. All the nodes work together to increase the efficiency and speed of the processors. A catalog ID number is assigned to each and every file, and is incremented when the file is added.

Also, we'll study some important concepts that play a crucial role in extraction:

Partition

Every iOS device has two partitions; one is the firmware partition, and the second partition is the user data partition. Initially, firmware partitions are read only; the user can't make any modifications with it, but after performing a firmware upgrade, it can be used as read and write as well. iTunes overwrites this partition with a new partition at the time of firmware upgrade. The size of this varies from 1 GB to 3 GB. Firmware partition does not have any user data. It just contains system upgrade and basic application files.

The second partition completely owns the user data. This is the partition in which we are going to run our investigations.

SQLite database

This is the most popular database used for mobile application development. This database is a relational database and is consolidated in a small C programming language. This is a mobile-compatible, lightweight database and supports all platforms. Many applications such as Notes, Images, Calendar schedules, SMS, and so on use this database in order to store and organize data. You can also open and view this database for investigation purposes. In the upcoming chapters, we are going to study this in detail.

How to do it...

We have read about the various tools to extract data in the past few chapters. Now, once the data is extracted from the device, you should be able to see files with multiple formats. Most of the files in the iOS device disk are in the PLIST format, that is, property list files. This is used to store data in COCOA and COCOA touch OS.

1. Launch any file with the `.plist` extension; you should be able to see the following format in Xcode, which is a development IDE for iOS application.

Key	Type	Value
▼ animations	Diction...	(4 items)
▼ knight-walk-left	Diction...	(2 items)
▼ frames	Array	(4 items)
Item 0	String	knight-walk-left01.png
Item 1	String	knight-walk-left02.png
Item 2	String	knight-walk-left03.png
Item 3	String	knight-walk-left04.png
delay	Number	0.2
▶ knight-walk-right	Diction...	(2 items)
▶ knight-attack-left	Diction...	(2 items)
▶ knight-attack-right	Diction...	(2 items)

2. Now, once you open the file in some other editor, you should be able to see this XML format to save the data on a key value basis.

3. This file can store multiple types of value such as strings, Boolean values, numbers, or binary values. This file is heavily used for saving in-app data by developers.

4. Few PLIST files are saved in a binary format, which will need special tools to decrypt them. **Plutil** is one of the open source tools that can be used on Mac/Windows/Linux for converting PLIST files into a human-readable format.

5. The following is the list of database files that can be used for specific information based on requirements.

File Name	Description
`AddressBook.sqlitedb`	This is contact information and personal data such as name, e-mail address, birthday, organization, and so on
`AddressBookImages.sqlitedb`	These are the images associated with saved contacts
`Calendar.sqlitedb`	This is calendar details and events information
`Call_history.db`	This is incoming and outgoing call logs including phone numbers and time stamps
`Sms.db`	This is text and multimedia messages along with their timestamps
`Voicemail.db`	This is voicemail messages
`Safari/Bookmarks.db`	This is saved URL addresses
`Safari/History.plist`	This is the user's internet browsing history
`Notes.sqlite`	This is Apple Notes application data
`Maps/History.plist`	This keeps a track of location searches
`Maps/Bookmarks.plist`	This is saved location searches
`consolidated.db`	This stores GPS tracking data
`En_GB-dynamic-text.dat`	This is the keyboard cache
`com.apple.accountsettings.plist`	This maintains data about all the e-mail accounts that are configured on the Apple Mail application
`com.apple.network.identification.plist`	This includes wireless network data including IP address, router IP address, SSID, and timestamps

6. You can see all these files inside the iPhone backup repository. This should look something similar to the following screenshot:

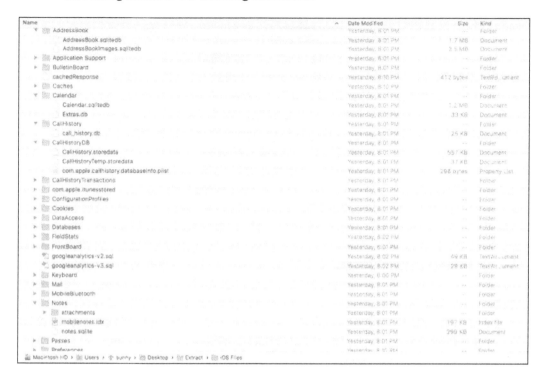

7. Other than the listed files, it also contains third-party application files. There are multiple encrypted third-party files, which can be decrypted for the purpose of investigation. Facebook, Google, and so on save their authentication tokens and cookies in the plist file, which can act as a backdoor entry for authentication on a user account.

8. The file encryption system on iOS looks something similar to the following screenshot:

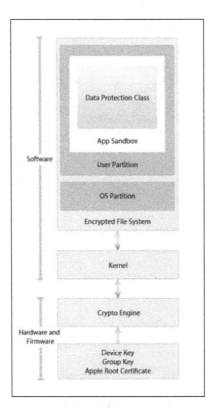

9. From iPhone 3GS onwards, Apple has started protecting user data by encrypting user partitions on the disk. They also use the device hardware key as the component for the encryption process, which also means that the physical device is needed while extracting information from the user's partition.

10. With the introduction of iOS 5, Apple has introduced one more layer of protection by encrypting files with individual keys. All the tools that we have studied in previous chapters allow us to decrypt data in raw files, which can be further read.

See more...

To explore in detail, visit the following link:

http://resources.infosecinstitute.com/iphone-forensics-2/

Recovering backups

The backup format for Apple devices is a bit different from other devices. To back up our Apple device we need to have iTunes installed on our computer and we can also take a backup on iCloud. This backup analytics is available from iOS 5 onwards.

How to do it...

Let's discuss how to take a backup on iTunes, and on iCloud as well.

iCloud backup:

iCloud is a kind of virtual data storage on the Internet. To store anything on iCloud, the user needs to register with an Apple account. On iCloud, the user can keep a backup of their photos, application data, contacts, messages, e-mail, and so on. To use the iCloud storage, the user needs to connect with the Internet. iCloud services don't depend on computers to store their data. It is a kind of computer-free backup solution. iCloud data can be managed and accessed by any of the Apple devices—iPhone, iPad, iPod, and computers as well. The user just needs to login with his Apple account credentials. iCloud allows a user to store data up to 5 GB without any charge; if the user wants more memory, then they can purchase storage from Apple. iCloud data is stored encrypted in their server using 128-bit encryption algorithm and they decrypt our data on the air when it is requested from the authenticated device.

Here are the steps to update iCloud settings:

1. On Apple devices, we can manually set the iCloud storage to On/Off by navigating to **Settings**.
2. Then, go to **iCloud**.
3. Tap on **Storage**.
4. Go to **Manage Storage**.
5. Then, hit **Backup** (click on the device name).

6. Toggle can be enabled/disabled for each application installed as shown in the following screenshot:

As mentioned here, data is stored in an encrypted manner so Apple claims that the user's data is safe from hackers. They also enforce users to use strong passwords.

iTunes backup:

While backing up your Apple devices, iTunes asks the user for the type of backup (it can be encrypted and decrypted). An unencrypted backup is completely accessible, while encrypted data is protected by the strong password given by the owner. Every time a backup is encrypted, it will be affiliated with the same password the user gave the first time. For this reason, if a password has already been set, during the forensic backup, we will get the encrypted backup or data.

The iTunes backup folders for different operating systems are as follows:

Operating system	iTunes backup folders
MAC OS X	`~/Library/Application Support/MobileSync/Backup/` (where "~" represents the user's home directory)
Windows XP	`\Documents and Settings\(username)\Application Data\Apple Computer\MobileSync\Backup\`
Windows Vista, Windows 7,8,10	`\Users\(username)\AppData\Roaming\Apple Computer\MobileSync\Backup\`

iTunes created a backup every time it synced. The backup contains a .plist file, which includes DSID, IMEI, serial number, and much more information about the iDevice; this file is called a Manifest file. This Manifest file can be located inside the backup folder with the name, `info.plist`.

You can locate the `manifest.plist` file using the following screenshot:

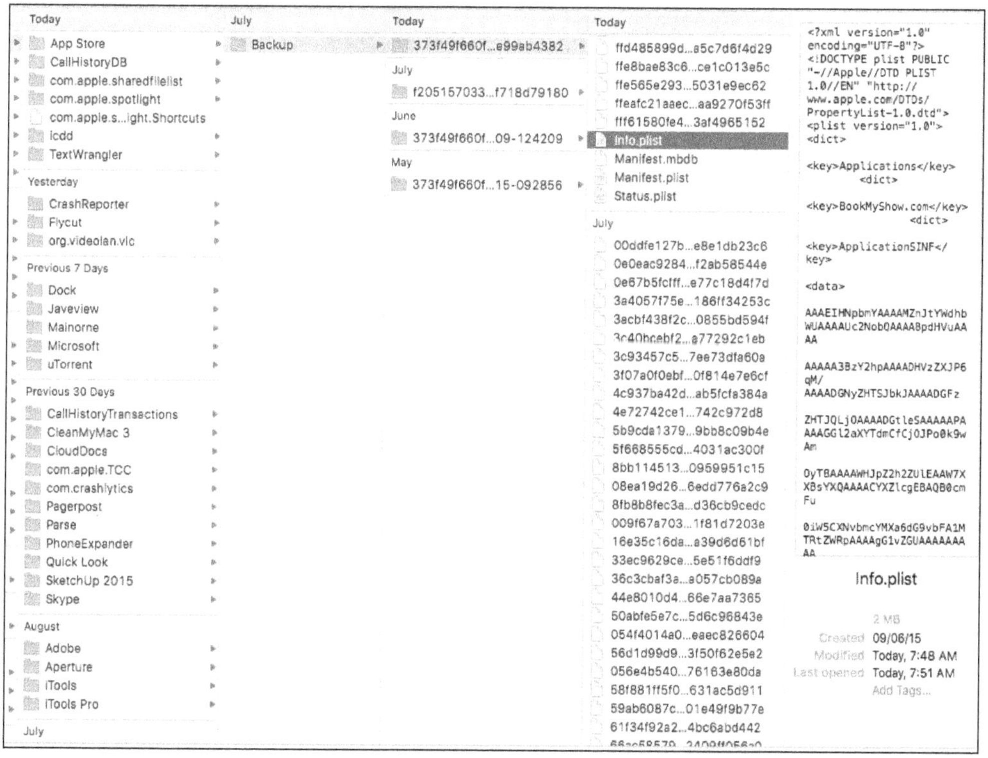

Here is a sample of the `manifest.plist` file:

```xml
<?xml version="1.0" encoding="UTF-8"?>
<!DOCTYPE plist PUBLIC "-//Apple//DTD PLIST 1.0//EN" "http://www.
apple.com/DTDs/PropertyList-1.0.dtd">
<plist version="1.0">
<dict>
  <key>Applications</key>
  <dict>Applications and Data </dict>
  <key>Build Version</key>
  <string>13B137</string>
  <key>Device Name</key>
  <string>iPhone</string>
  <key>Display Name</key>
  <string> iPhone </string>
  <key>GUID</key>
  <string>268540233E1029AF04CC32EA1C666CF3</string>
  <key>ICCID</key>
  <string>89914902900000436861</string>
  <key>IMEI</key>
  <string>990002256303189</string>
  <key>Installed Applications</key>
  <array>List of Installed Applications </array>
  <key>Last Backup Date</key>
  <date>2015-10-16T02:18:07Z</date>
  <key>MEID</key>
  <string>99000225630318</string>
  <key>Phone Number</key>
  <string>1234567890</string>
  <key>Product Name</key>
  <string>iPhone 5</string>
  <key>Product Type</key>
  <string>iPhone5,2</string>
  <key>Product Version</key>
  <string>9.1</string>
  <key>Serial Number</key>
  <string>DNPJFALCF8GH</string>
  <key>Target Identifier</key>
  <string>373f49f660f4b7e296601b9c6a5b8d8e99ab4678</string>
  <key>Target Type</key>
  <string>Device</string>
  <key>Unique Identifier</key>
  <string>373F49F660F4B7E296601B9C6A5B8D8E99AB4678</string>
  ...
```

Let's discuss the code in detail as follows:

1. iTunes stores the data in a separate folder for each device.

2. Inside this folder, there is a subfolder for each device. The name of the folder is equivalent to the UDID of the device, which contains a 40-character hexadecimal string.

3. iTunes holds a single backup for each device and adds only those files to which a modification has been made or if they are newly added.

4. When an iDevice is connected to iTunes, it start recognizing the device if it is already authorized; then, it starts the backup and syncs with the machine without asking for the passcode as it is already authorized with that machine.

5. While backing up, iTunes creates a plist file with the name UDID to store some info such as Host ID, Host private key, Device, the host certificate, and so on.

6. The Escrow Keybag is used to give full access to the paired device, which is currently connected when the device is in a locked state. It improves the efficiency and usability for the user and it won't ask the user to unlock the device; it handles it in the locked state as well. Escrow Keybag locates different OSes on different locations and the accurate paths are listed in the next section.

Key generation

A unique key is generated from the iDevice hardware (key 0 x 835), which is used to encrypt the Escrow Keybag. It is secured with a passcode of 32 bytes and it is stored in the iDevice as well.

The accurate paths are listed in the following table:

Operating system	Escrow Keybag Location
MAC OS X	`/private/var/db/lockdown/`
Windows	`%AllUsersProfile%\Apple\Lockdown\`

To prevent attacks from hackers from iOS 5 onwards, Apple uses the Escrow Keybag by protecting it with the user's passcode as well. Escrow Keybag is stored in a PLIST file in the `escrow_records` directory by navigating to /private/var/root/Library/Lockdown. Escrow Keybag contains all the keys of the protection classes that are made for data encryption. iTunes allows users to access the keys stored in the Escrow Keybag even if the iPhone is locked.

Here are the steps to create encrypted backups of an iDevice from a Mac:

1. Connect the iDevice to you Mac and open iTunes.

2. The user will be prompted for the password while encrypting the backup.

3. In the future, we will need this password to encrypt all the files. iTunes also uses it to save the given credentials in Keychain.

4. Keybag is itself encrypted with the backup password, through which we can decrypt it even without having physical access. This should look similar to the following screenshot:

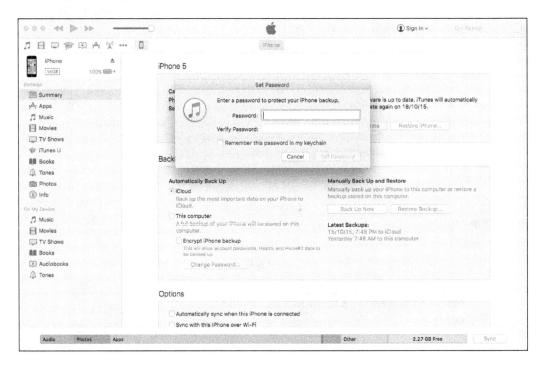

5. iTunes takes a backup of everything, such as music, call logs, keychain, web cache, application data, SMS, chat history, contacts, and so on. It also backs up the attached device information.

6. Files from the backup are not in a readable state because of encryption and are named with a 40-digit random, alphanumeric hex value with a lack of extension as well.

7. This 40-digit hex file name is appended with the respective domain name with this "-" prefix. This should look similar to the following screenshot:

See more...

To explore this, visit the following links:

- `http://resources.infosecinstitute.com/ios-5-backups-part-1/`
- `http://forums.blipinteractive.co.uk/node/2412`
- `http://www.zdziarski.com/blog/wp-content/uploads/2013/05/iOS-Forensic-Investigative-Methods.pdf`

Extracting data from iTunes backups

Extracting the logical information from the iTunes backup is crucial for forensics investigation. There is a full stack of tools available for extracting data from the iTunes backup. They come in a wide variety, from open source to paid tools. Some of these forensic tools are Oxygen Forensics Suite, Access Data MPE+, EnCase, iBackup Bot, DiskAid, and so on. The famous open source tools are iPhone Backup Analyzer and iPhone Analyzer. In this section, we are going to learn how to use the iPhone backup extractor tools.

How to do it...

The iPhone backup extractor is an open source forensic tool that can extract information from device backups. However, there is one constraint that the backup should be created from iTunes 10 onwards. Follow these steps to extract data from iTunes backup:

1. Download the iPhone backup extractor from `http://supercrazyawesome.com/`.

2. Make sure that all your iTunes backup is located at this directory: `~/Library/ApplicationSupports/MobileSync/Backup`. In case you don't have the required backup at this location, you can also copy and paste it.

3. The application will prompt after it is launched. The prompt should look similar to the following screenshot:

4. Now tap on the **Read Backups** button to read the backup available at `~/Library/ApplicationSupports/MobileSync/Backup`. Now, you can choose any option as shown here:

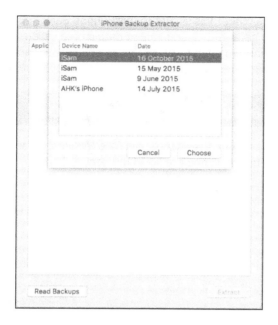

5. This tool also allows you to extract data for an individual application and enables you to read the iOS file system backup.

6. Now, you can select the file you want to extract. Once the file is selected, click on **Extract**.

7. You will get a popup asking for the destination directory. This complete process should look similar to the following screenshot:

8. There are various other tools similar to this; iPhone Backup Browser is one of them, where you can view your decrypted data stored in your backup files. This tool supports only Windows operating systems as of now. You can download this software from `http://code.google.com/p/iphonebackupbrowser/`.

Encrypting and decrypting tools

Another backup format came into the picture using the Manifest file with the extension `.abdb`. To retrieve these backups, find the file in the backup folder.

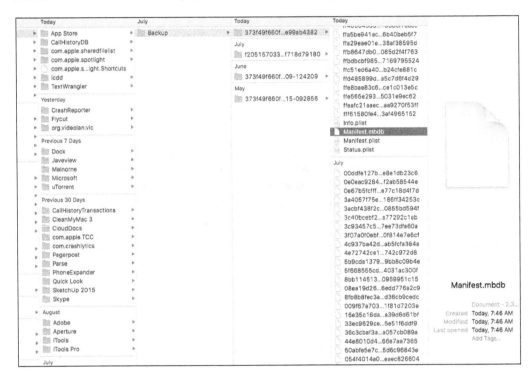

The Manifests uses a proper binary format. Nowadays, in open source, plenty of scripts are available to parse the data.

How to do it...

1. A sample for the Python script to read the Manifest is as follows:

```
#!/usr/bin/env python
import sys
import shutil
import os
```

```
import errno

def mkdir_p(path):
try:
os.makedirs(path)
except OSError as exc: # Python >2.5
if exc.errno == errno.EEXIST:
pass
else: raise

def getint(data, offset, intsize):
"""Retrieve an int (big-endian) and new offset from the current
offset"""
value = 0
while intsize > 0:
value = (value<<8) + ord(data[offset])
offset = offset + 1
intsize = intsize - 1
return value, offset

def getstring(data, offset):
"""Retrieve a string and new offset from the current offset into
the data"""
if data[offset] == chr(0xFF) and data[offset+1] == chr(0xFF):
return '', offset+2 # Blank string
length, offset = getint(data, offset, 2) # 2-byte length
value = data[offset:offset+length]
return value, (offset + length)

def process_mbdb_file(filename):
mbdb = {} # Map offset of info in this file => file info
data = open(filename).read()
if data[0:4] != "mbdb": raise Exception("Not an MBDB file")
offset = 4
offset = offset + 2 # value x05 x00, not sure what this is
while offset < len(data):
fileinfo = {}
fileinfo['start_offset'] = offset
fileinfo['domain'], offset = getstring(data, offset)
fileinfo['filename'], offset = getstring(data, offset)
fileinfo['linktarget'], offset = getstring(data, offset)
fileinfo['datahash'], offset = getstring(data, offset)
fileinfo['unknown1'], offset = getstring(data, offset)
fileinfo['mode'], offset = getint(data, offset, 2)
fileinfo['unknown2'], offset = getint(data, offset, 4)
fileinfo['unknown3'], offset = getint(data, offset, 4)
fileinfo['userid'], offset = getint(data, offset, 4)
fileinfo['groupid'], offset = getint(data, offset, 4)
fileinfo['mtime'], offset = getint(data, offset, 4)
fileinfo['atime'], offset = getint(data, offset, 4)
```

```
fileinfo['ctime'], offset = getint(data, offset, 4)
fileinfo['filelen'], offset = getint(data, offset, 8)
fileinfo['flag'], offset = getint(data, offset, 1)
fileinfo['numprops'], offset = getint(data, offset, 1)
fileinfo['properties'] = {}
for ii in range(fileinfo['numprops']):
propname, offset = getstring(data, offset)
propval, offset = getstring(data, offset)
fileinfo['properties'][propname] = propval
mbdb[fileinfo['start_offset']] = fileinfo
return mbdb
def process_mbdx_file(filename):
mbdx = {}
data = open(filename).read()
if data[0:4] != "mbdx": raise Exception("Not an MBDX file")
offset = 4
offset = offset + 2 # value 0x02 0x00, not sure what this is
filecount, offset = getint(data, offset, 4) # 4-byte count of
records
while offset < len(data):
# 26 byte record, made up of ...
fileID = data[offset:offset+20] # 20 bytes of fileID
fileID_string = ''.join(['%02x' % ord(b) for b in fileID])
offset = offset + 20
mbdb_offset, offset = getint(data, offset, 4) # 4-byte offset
field
mbdb_offset = mbdb_offset + 6 # Add 6 to get past prolog
mode, offset = getint(data, offset, 2) # 2-byte mode field
mbdx[mbdb_offset] = fileID_string
return mbdx

def extract_file(f, verbose=False):
print "Processing %s::%s" % (f['domain'], f['filename'])
if len(f['filename']) > 0:
path = f['domain'] + "/" + f['filename']
    c = path.split("/")
    c.pop()
        dirname = './filesystem/' + '/'.join(c)
    mkdir_p(dirname)
try:
shutil.copy2(f['fileID'], './filesystem/%s'
%(path))
except:
pass
verbose = True
if __name__ == '__main__':
mbdb = process_mbdb_file("Manifest.mbdb")
mbdx = process_mbdx_file("Manifest.mbdx")
for offset, fileinfo in mbdb.items():
if offset in mbdx:
```

```
fileinfo['fileID'] = mbdx[offset]
else:
fileinfo['fileID'] = "<nofileID>"
print >> sys.stderr, "No ID found for %s" %
extract_file(fileinfo)
extract_file(fileinfo, verbose)
```

2. To execute this script, change the directory to the backup folder you'd like to extract, then run Python with the pathname to the preceding script:

    `$ python dump_mbdb_10.py`

3. A new folder named `filesystem` will be created in your current working directory containing the reconstructed backup.

4. This script extracts files into an order of domains; application domains contain data of installed applications on our device, such as media (this contains images, videos, and other things) and other domains.

5. An iTunes backup mainly includes three property lists, which keep the record of important information. Go to your finder inside the backup directory. You can see the long list of backup files. Locate **info.Plist**:

The following Key values are seen under **info.plist**:

- **Applications**: This contains a list of applications installed on the device.

- **Device Name**: This holds the record of the Owner name of the device.

- **ICCID**: This contains the serial number of the SIM.

- **IMEI**: This holds the IMEI number of the device.

- **Last Backup Date**: This is the date when the last backup was taken.

- **Phone Number**: This records the phone number if it is a cellular phone while taking the backup.

- **Product Name**: This is the name of the product (either iPhone/iPad).

- **Product Type**: This is the type of the product with the version.

- **Product Version**: This is the product version or software version of the device at the time of taking the backup.

- **Serial Number**: This is the serial number of the product that has been backed up.

- **iTunes Settings**: This holds the list of applications installed on the devices. It also contains the list of synced applications that are already synced to the machine.

- **iTunes Version**: This holds the version of the iTunes that created the backup the last time.

- **Unique Identifier** and **Target Identifier**: They are the unique IDs of iDevice (**DSID**). Through this ID pairing between the machines, the devices are very secure.

6. Now, open the `Manifest.plist` file, which should look similar to the following screenshot:

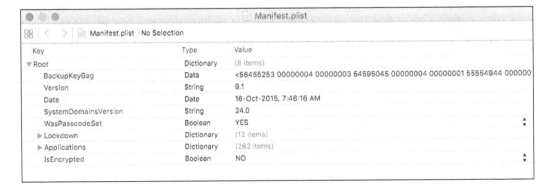

We can retrieve the following details from this file:

- **Date**: This stores the time at which the backup was taken or updated
- **WasPasscodeSet**: The name itself identifies that it tracks a record whether the passcode was set at the time of syncing
- **Lockdown**: This keeps a record of the `com.apple.mobile.data_sync` profile, which in the future identifies the sources to which we configured the devices
- **Applications**: This handles all the installed application inner devices and keeps a track of the applications such as their version number, and so on
- **IsEncrypted**: This identifies whether the backup is encrypted

7. Open the `Status.plist` file. This file should look similar to the following screenshot:

The following are the details of the file:

- **IsFullBackup**: This identifies whether the backup of the iDevice was taken fully
- **BackupState**: This state identifies whether it's a new backup or already has been updated
- **Date**: This stores the time at which the backup was taken or updated

Manifest.mbdb

`Manifest.mbdb` is the file that manages the backup scenario for Apple devices. This file holds the information of all the other files such as their structure, sizes, indexes, and so on.

In earlier versions of OS and iTunes, the backup file structure was supervised by two different files: `.mbdx` and `.mbdb`. Here, the index for the backup is handled by `.mbdx` and the element of those indexes can be located in `.mbdb`. From iTunes 10 onwards, Apple discarded index files and all the backup is maintained by a single `.abdb` file.

A sample Manifest file is shown in the following screenshot. As a Manifest file is a binary file, a Hex editor is accommodated to read the content.

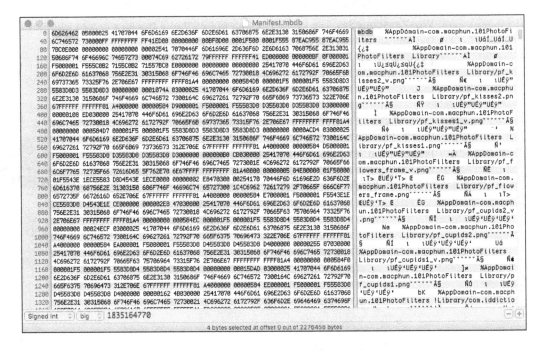

The `Manifest.mbdb` file header and record format are discussed as follows:

Header: The `.mbdb` file header is a fixed value of 6 bytes and the value is the number to identify the `.mbdb` files; its value is as follows:

Type	Value
uint8[6]	mbdb\5\0

Record: The `.mbdb` file contains many records and each record is of a variable size. Every record contains various details about a file.

Type	Data	Description
string	Domain	This is the domain name
string	Path	This is the file path
string	Target	This is the absolute path for symbolic links
string	Digest	It works on the SHA 1 hash function. The values obtained are mostly none (0xff 0xff) for Directories and AppDomain Files and 0x00 0x14 for System Domain files

Type	Data	Description
string	Encryption key	The value is None (0xff 0xff) for un encrypted files
uint16	Mode	This record identifies the File Type and the types described are as follows: '0xa000' for a symbolic link '0x4000' for a directory '0x8000' for a regular file
uint64	inode number	It gives a lookup entry in the inode table
uint32	User ID	Mostly 501
uint32	Group ID	Mostly 501
uint32	Last modified time	It shows a file's last modified time in Epoch format
uint32	Last accessed time	It shows a file's last accessed time in Epoch format
uint32	Created time	It shows a file's creation time in Epoch format
uint64	Size	Length of file '0' for a symbolic link '0' for a directory Non zero for a regular file
uint8	Protection class	Data protection class (values 0x1 to 0xB)
uint8	Number of Properties	This is the number of properties

In the backup, most of the information is stored as plist files, SQLite database files, and image files. Backup files can be viewed directly by adding an appropriate file extension.

For example, adding a .plist file extension to the `bc0e135b1c68521fa4710e3edadd6e74364fc50a` file allows viewing the contents of the Skype property list file using a plist editor.

There are many free tools available to read iTunes backups. Some well-known tools are listed here:

- On MAC OS X, there is iPhone Backup Extractor (`http://supercrazyawesome.com/`)

- On Windows, there is iPhone Backup Browser (`http://code.google.com/p/iphonebackupbrowser/`)

- On Mac OS X and Windows there is iBackupBot (`http://www.icopybot.com/itunes-backup-manager.htm`)

These tools parse the information stored in the `.mbdb` file and create the file structure. The tools convert the gibberish backup files into a readable format as shown in the following figure:

▼ Library	Today, 9:14 PM	--	Folder
▶ Accounts	Today, 8:01 PM	--	Folder
▼ AddressBook	Today, 8:01 PM	--	Folder
AddressBook.sqlitedb	Today, 8:01 PM	1.7 MB	Document
AddressBookImages.sqlitedb	Today, 8:01 PM	3.5 MB	Document
▶ Application Support	Today, 8:01 PM	--	Folder
▶ BulletinBoard	Today, 8:01 PM	--	Folder
cachedResponse	Today, 8:10 PM	412 bytes	TextEd...ument
▶ Caches	Today, 8:10 PM	--	Folder
▶ Calendar	Today, 8:01 PM	--	Folder
▶ CallHistory	Today, 8:01 PM	--	Folder
▶ CallHistoryDB	Today, 8:01 PM	--	Folder
▶ CallHistoryTransactions	Today, 8:01 PM	--	Folder
▶ com.apple.itunesstored	Today, 8:01 PM	--	Folder
▶ ConfigurationProfiles	Today, 8:01 PM	--	Folder
▶ Cookies	Today, 8:01 PM	--	Folder
▶ DataAccess	Today, 8:01 PM	--	Folder
▶ Databases	Today, 8:01 PM	--	Folder
▶ FieldStats	Today, 8:02 PM	--	Folder
▶ FrontBoard	Today, 8:01 PM	--	Folder
googleanalytics-v2.sql	Today, 8:02 PM	49 KB	TextWr...ument
googleanalytics-v3.sql	Today, 8:02 PM	29 KB	TextWr...ument
▶ Keyboard	Today, 8:00 PM	--	Folder
▶ Mail	Today, 8:01 PM	--	Folder
▶ MobileBluetooth	Today, 8:01 PM	--	Folder
▶ Notes	Today, 8:01 PM	--	Folder

Source: http://resources.infosecinstitute.com/ios-5-backups-part-1/

These tools use the iDevice API to read/write in backups. The scope of backup extractors is limited; they can only read the non-protected files because protected files are encrypted.

For example, the backup extractor provides us with **keychain-backup.plist**, which can be opened in the plist editor. But still, the protected/encrypted data will not be in a readable form, as shown in the following screenshot:

With iOS 4, Apple introduced a better data protection mechanism through which we can protect our crucial data or credentials and keychain items by one more protection layer. In this process, along with the device-specific hardware encryption, the user passcode is also used to make a set of class keys, which sets a protection layer on the data. Data Protection is used by developers to add one more layer to the security for files and keychain items. Protection class keys are stored in System Keybag, on the iDevice itself. A new set of protection class keys are generated and stored in the Backup Keybag while taking a backup every time.

To enable Data Protection, the `setAttributes:ofItemAtPath:error` method is available in the `NSFileManager` class. It can be set by giving the `NSFileProtection` attribute. The following are the protection classes for the files:

Key id	Protection class	Description
1	`NSProtectionComplete`	The file is accessible only after the device is unlocked
2	`NSFileProtectionCompleteUnlessOpen`	The file is accessible after the device is unlocked OR The file is accessible if the file handle remains open before locking the device
3	`NSFileProtectionCompleteUntilFirstUser Authentication`	The file is accessible after the first unlock of the device till reboot
4	`NSProtectionNone`	The file is accessible even if the device is locked
5	`NSFileProtectionRecovery`	This is undocumented

Data protection for keychain items can be enabled by setting a protection class value in the `SecItemAdd` or `SecItemUpdate` methods. Keychain class keys also define whether a keychain item can be migrated to another device or not. A list of protection classes available for keychain items is shown in the following table:

Key id	Protection class	Description
6	`kSecAttrAccessibleWhenUnlocked`	The keychain item is accessible only after the device is unlocked
7	`kSecAttrAccessibleAfterFirstUnlock`	The keychain item is accessible only after the first unlock of the device to till reboot
8	`kSecAttrAccessibleAlways`	The keychain item is accessible even when the device is locked

Key id	Protection class	Description
9	`kSecAttrAccessibleWhenUnlockedThis DeviceOnly`	The keychain item is accessible only after the device is unlocked and the item cannot be migrated between devices
10	`kSecAttrAccessibleAfterFirstUnlock ThisDeviceOnly`	The keychain item is accessible after the first unlock of the device and the item cannot be migrated
11	`kSecAttrAccessibleAlwaysThisDeviceOnly`	The keychain item is accessible even when the device is locked and the item cannot be migrated

Source: http://securitylearn.net/wp-content/uploads/iOS%20Resources/Forensic%20analysis%20of%20iPhone%20backups.pdf

Decrypting iTunes backups

iTunes gives the user the option to encrypt their backup. The passcode is set in the device itself, while backing up the Apple device used to send the encrypted file or data to iTunes.

It works as an extra layer of security where, without this password, we can't recover the encrypted data from the iTunes backup. If you locate an encrypted backup from the machine and have decrypted the keychain data then, follow the following steps to decrypt the backup data.

1. **Locating the encryption password**: The tool named **Recover-keys.sh** is used to decrypt the keys from the iDevice, along with the key script, also decrypted and saved into the file with the prefix "keychain-" and with the extension `.txt`. Locate the `BackupPassword` file in decrypted files. This will be associated as your iTunes backup encryption password. Save it somewhere; we will need it later as our credentials.

2. **Downloading and installing Python and PyCrypto**: Now, download and install Python and PyCrypto. We need the Python interpreter for scripts written in Python, as the iTunes decryption utility. Download and install Python from `http://www.python.org`.

We can install Python by the following MacPorts command:

```
$ sudo port install python26
```

After installing Python, we need to download PyCrypto. It is used as a cryptographic utility library. So download it from the following link and install it:

`http://www.pycrypto.org`

Extract, build, and install the PyCrypto module as follows:

```
$ tar -zxvf pycrypto-2.3.tar.gz
$ cd pycrypto-2.3
$ python2.6 setup.py build
$ sudo python2.6 setup.py install
```

3. **Downloading and running the Decryption utility**: Locate the file named `mbdb_10.zip` from the repository's `Scripts` directory, and download it. Unarchive the ZIP file. It consists of a number of Python classes and Python scripts, and they are used to decrypt the backup from iTunes.

   Run the `decrypt_mbdb_10` script by providing the following details in this order—encrypted backup directory path, a target path to decrypt to, and the `BackupPassword` key:

   ```
   $ cd decrypt_mbdb_10
   $ python2.6 decrypt_mbdb_10.py
   3C333086522b0ea392f686b7ad9b5923285a66af decrypted PASSWORD
   ```

Now, you can locate the script file, which is being run. Meanwhile, the console messages will update you about the progress regularly. After the script completes, you can locate the number of domain directories within the given decrypted path.

See more...

▶ Refer `https://www.exploit-db.com/docs/19767.pdf` to explore more information on this topic

7
Forensics Tools

In this chapter, we will be focusing on the following recipes:

- ▶ Exploring iPhone Backup Analyzer
- ▶ Exploring iExplorer
- ▶ Exploring SQLite browser
- ▶ Jailbreaking iPhone devices

In this chapter, we are going to learn about some advanced forensic tools. This will include various techniques of Jailbreaking iPhone devices. We will also explore the SQLite browser, which can be used to recover the database of the applications installed on Apple devices after recovering files from the recovery tools.

Exploring iPhone Backup Analyzer

In this section, we are going to explore the iOS forensic tool, iPhone Backup Analyzer.

Getting ready

iPhone Backup Analyzer is a Java-based product that can be used on multiple platforms, such as Windows, Linux, and Mac. You can download the JAR file from `http://sourceforge.net/projects/iphoneanalyzer/files/latest/download`. This is an open source tool that can be used to extract the iTunes backup files. Most of the time, this tool is used for extracting and analyzing existing backups. However, if needed, this tool can also make a backup.

How to do it...

Perform the following steps to analyze the iTunes backup:

1. The first step is to get the copy of the backup data; in order to locate the data, you can go to the following directory:

    1. For Windows 7 and above: `<user_home>\Application Data\Apple Computer\MobileSync\Backup`

    2. For Mac: `/Library/Application Support/MobileSync/Backup`

2. Once you have navigated to the preceding path, you should be able to see something like the following screenshot:

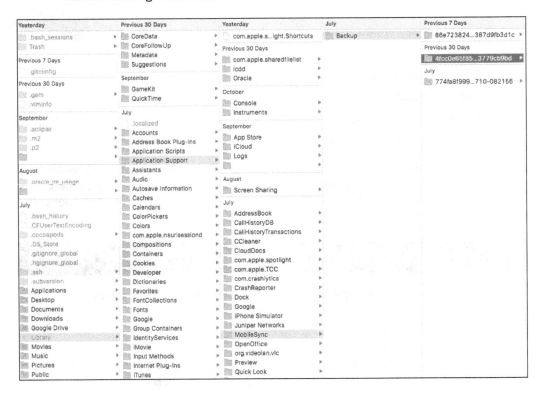

3. The directory can have more than one backup file. Each will have a name coded in a 40-digit hexadecimal value as shown in the following screenshot:

4. Now, once you take a deeper look into the directory files, you will see a huge number of files again with the same 40-digit hexadecimal filenames. Each file is used by the iOS device to provide you seamless access to data. These files are encrypted by iTunes while creating the backup and will be decrypted while reading.

5. Despite all the encoded files, there are few raw files in the directory. If you scroll down to bottom, you should be able to find `info.plist`, Manifest files, and `Status.plist`. This should look like the following screenshot:

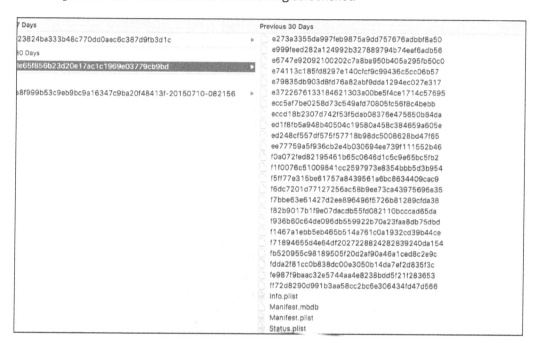

6. These files have a lot of information related to the backup date, times, and data available and stored; it also has data format information. Apart from all these details, these files also hold device information and iOS version details. These are the details used by the iPhone Backup Analyzer tool.

Now, since we have an Apple device's backup ready, we can run our iPhone Analyzer.

1. Open **iPhone Analyzer** and click on the **File** menu. This should look like the following screenshot:

2. This tool allows you to open the existing backup located at different paths from the original iTunes directory.

3. You can also open the backups created by iTunes by clicking on **Open: Default iTunes location**.

4. Once you select any option from the file menu, you will be redirected to the file browser, where you can select the backup to analyze.

5. Once the backup is selected, you should be able to see the screen that has multiple options to read the data. On the right-hand side of the window, you should be able to see **Phone Information**. Clicking on it will show you detailed information about the device. This should look something similar to following screenshot:

6. The information shown in the preceding screenshot is extracted from the Info.plist file we saw while exploring the backup directory.

7. The most significant information is located in the first section of the window that includes your device serial number, phone number, device name, and so on. The second section mostly covers granular details and information. This can be in the form of files saved by various apps, device settings, and so on.

8. You can also find the `Manifest` file in the backup tool; a few files' availability depends on the version of the operating system. They are also not always human-readable files; however, using the tool you can link and read the files, as shown in the following screenshot:

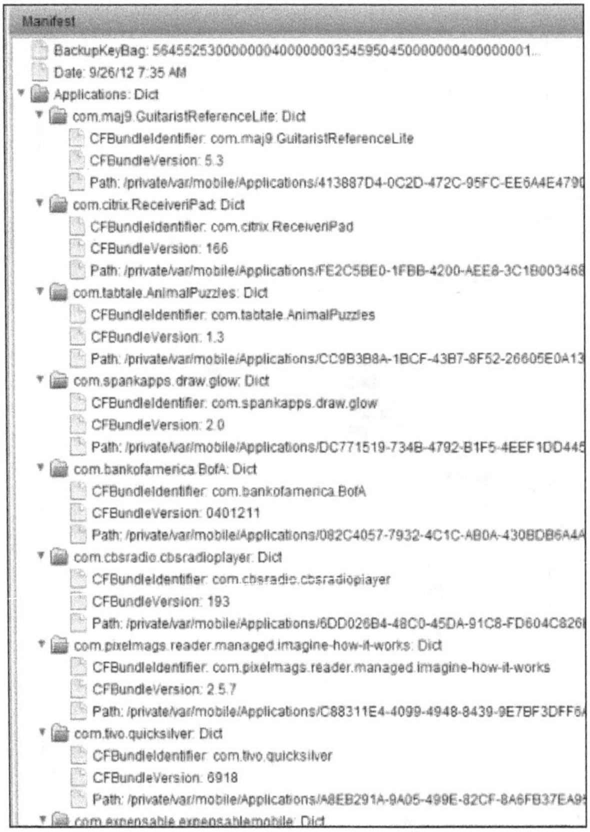

9. Now, since we have got a chance to get our hands on the basic device information, we will move forward with more detailed information recovery methods. On the left-hand side of the tool, you should be able to see two tab bookmarks and file systems. It should look like the red highlighted area in the following screenshot:

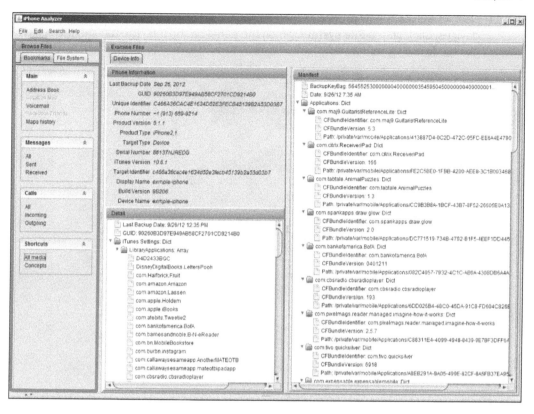

10. Bookmarks make it easy to explore the device, as all the shortcuts come with the tool. They include all the frequently visited tabs and are arranged accordingly. Bookmarks consist of the following:

- **Address book**: This will allow you to explore and recover all the contact details.

- **Map history**: This includes detail of all the recent GSM and Wi-Fi access points in order to fetch the geolocation history of the device and its whereabouts.

- **Voicemail**: Here, you can find the metadata and sound clips of all your recent voicemails.

- **Messages**: Explore all the sent and received messages using this tab.

- **Calls**: By diving in deep with this feature, you can see the list of all the incoming, outgoing, and missed calls.

- **Shortcuts**: Here, you can select **All media** to take a look at all the images and media files stored in the device.

11. Now we will learn how to explore the File system and device directories of Apple devices. Click on the **File system** tab on the panel on the left. The tab should look like the following screenshot:

12. There will be two types of structure in the File system. One will be the directory and the second will be the files. You can always parse the complete tree of directories in order to explore the files and content details of it.

13. The file system is divided into the following four parts:

- Documents
- Library
- Media
- System configuration

14. Most of the files we explore will have SQLite databases. There are multiple native and consumer-based apps that use SQLite as the primary database. These files can be easily explored and opened in SQLite Browser.

15. SQLite Browser is an open source tool that allows us to read the data and content of the SQLite database files. You can write and customize queries in order to get relevant results from the application database.

16. Now we will see how to explore raw files in iPhone Analyzer. When you select a file either from the filesystem or from bookmarks, it will open in a new tab on the right side of the window.

17. This right panel has a tab-based structure; every new file will open in a new tab retaining the previous file. You can navigate back and forth in between files.

18. iPhone Analyzer allows you to explore files in a variety of formats. These are as follows:

   ❏ **Special view**: This is specifically designed for the type of file being opened. It switches the window based on the type of file.

   ❏ **Text**: This will open the file in the text format, where you can see the data in raw text.

   ❏ **Hex**: This will display the file in the raw hex format.

   ❏ **SQLite**: This view will show the data in the SQL format; this will arrange the data in a tabular format for easy investigation purposes.

   ❏ **Plist**: All the application bundles have files with the plist format, which can be read in this tab.

19. Once the file is selected, it should autocreate the formats in which it can be opened. This should look like the following screenshot:

20. iPhone Analyzer also provides a very interesting feature where we can collaborate on the data for group analysis. This data provides a detailed insight of correlation between various activities that happened on the phone. This feature is called as **Concept Explorer**. Once selected, this should look like the following screenshot:

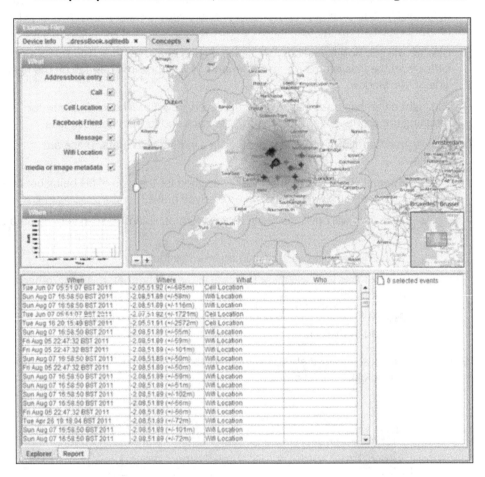

21. As shown in the preceding screenshot, we can see a new tab called **Concepts**, which has detailed information generated with a correlation of various entities. We can select and unselect these entities from the **What** section, which is located on the top left of the window. This looks similar to the following screenshot. All the data generated on the map is generated with the combined details of these entities:

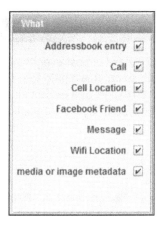

22. Apart from the map, this tool also provides data in a tabular format, which displays the date, location, and information about the network the data was generated from. This allows us to know whether the data has been generated from the mobile network or from any specific access point network. The tabular data should look similar to the following:

When	Where	What	Who
Tue Jun 07 05:51:07 BST 2011	-2.05,51.92 (+/-685m)	Cell Location	
Sun Aug 07 16:58:50 BST 2011	-2.08,51.89 (+/-58m)	Wifi Location	
Sun Aug 07 16:58:50 BST 2011	-2.08,51.89 (+/-116m)	Wifi Location	
Tue Jun 07 05:51:07 BST 2011	-2.07,51.92 (+/-1721m)	Cell Location	
Tue Aug 16 20:15:49 BST 2011	-2.05,51.91 (+/-2572m)	Cell Location	
Sun Aug 07 16:58:50 BST 2011	-2.08,51.89 (+/-55m)	Wifi Location	
Fri Aug 05 22:47:32 BST 2011	-2.08,51.89 (+/-59m)	Wifi Location	
Fri Aug 05 22:47:32 BST 2011	-2.08,51.89 (+/-101m)	Wifi Location	
Sun Aug 07 16:58:50 BST 2011	-2.08,51.89 (+/-50m)	Wifi Location	
Fri Aug 05 22:47:32 BST 2011	-2.08,51.89 (+/-50m)	Wifi Location	
Sun Aug 07 16:58:50 BST 2011	-2.08,51.89 (+/-59m)	Wifi Location	
Sun Aug 07 16:58:50 BST 2011	-2.08,51.89 (+/-51m)	Wifi Location	
Sun Aug 07 16:58:50 BST 2011	-2.08,51.89 (+/-102m)	Wifi Location	
Sun Aug 07 16:58:50 BST 2011	-2.08,51.89 (+/-56m)	Wifi Location	
Fri Aug 05 22:47:32 BST 2011	-2.08,51.89 (+/-66m)	Wifi Location	
Tue Apr 26 19:18:04 BST 2011	-2.08,51.89 (+/-72m)	Wifi Location	
Sun Aug 07 16:58:50 BST 2011	-2.08,51.89 (+/-101m)	Wifi Location	
Sun Aug 07 16:58:50 BST 2011	-2.08,51.89 (+/-72m)	Wifi Location	

| Explorer | Report |

23. There are various ways to extract the data that can be explored independently. This tool lets you access and analyze every file stored in the backup.

Exploring iExplorer

iExplorer is a tool that allows you to access almost everything on the device. You can migrate music, photos, messages, apps, and so on. It is a paid tool, and it allows you to analyze your device and the data in-depth. You can export and extract almost every sort of file from the device; it also allows you hassle-free access to all your application files where the data can be read.

How to do it...

Perform the following steps to explore devices using the iExplorer tool:

1. Download the latest iExplorer version for Mac from `https://www.macroplant.com/iexplorer/download-ie3-mac`.

2. Now go to your `Downloads` folder and initiate the installation process.

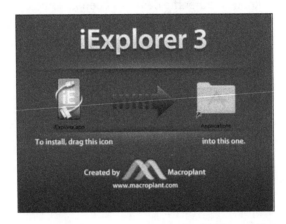

3. Then, drag iExplorer to the `Applications` folder. You can also pin the application on your dock. The iExplorer icon should look similar to the following screenshot:

4. Now, our installation process is completed. We will now see how to extract and browse our iTunes backups. This tool has a seamless way to parse your old iTunes backups, which includes your contacts, databases, photos, messages, notes, and so on.

5. Open iExplorer and connect your Apple device using a USB cable. You will be prompted by iTunes for sync, once the device is connected. Click on **Cancel**. It is recommended to not sync data using iTunes, as it might result in permanent data loss.

6. Navigate back to iExplorer, and it should show the following overview screen:

7. The preceding screen can be launched by clicking on the device name on the top left. In order to access the device backup data, click on **Backups**, which is highlighted in green in the preceding screenshot.

8. Old backups can also be accessed by clicking on **Backups** on the panel to the left.

9. If there is no backup on the device, then you will be prompted to take a backup, or you will be asked to use the last backup.

10. To browse the existing backup without connecting the device, click on **Explore iTunes Backups**.

11. Once you are in the section, you should be able to see **Browse iTunes Backup**. This section should list all your existing iTunes backups.

12. Select any backup from the list and click on **Backup Explorer**. This should allow you to access details such as messages, contacts, call history, and so on:

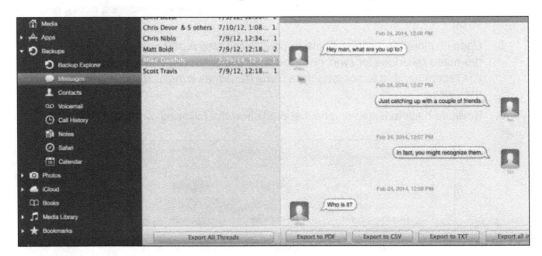

13. This also allows you to drag and drop the database files to your machine, and an export option is also available for messages, call history, and other similar features. You can export in multiple formats, such as PDF, CSV, and so on.

14. Now, in order to export the call history from the connected device, you can click on the **Data** tab.

15. In the **Data** tab, click on the **Call History** button. This should look like the following screenshot:

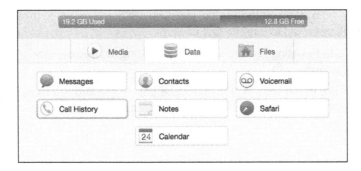

16. Once you click on **Call History**, it should give you the complete call history of the device. **Call History** will allow you to explore the call details made in that particular backup file. You can access the contact name, call duration, phone number, date time, and so on.

17. You can use a similar way to extract the call history from your iTunes backup files as well.

18. Just navigate to **Backup Explorer** and select the backup to explore. This should navigate you to a similar screen with options to explore the call history for that particular backup file. iExplorer also allows you to export the call details in multiple file formats.

There are various cases where we actually want to mount the Apple device as a disk. To mount the device, connect the device using a USB, open iExplorer, and perform the following set of steps:

1. First, navigate to **Media or Apps** in the left-hand side menu and right-click on **Media Library,** as shown in the following screenshot:

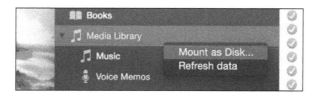

2. On clicking on **Mount as Disk...**, you will be prompted with a confirmation box. Click on **Okay** in the confirmation box.

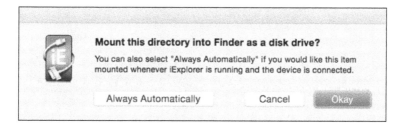

3. Now, once the confirmation is completed, open **Finder** and there you should be able to see your device details mounted as a disk.

4. This will allow you to explore your device at the file browser level.

Exploring SQLite browser

In this section, we are going to explore the way to recover and check data stored in SQLite database. This tool is going to help us to analyze and read SQLite data once it is recovered using the forensic recovery tools we used in the preceding sections.

Getting ready

Download SQLite Browser from `http://sqlitebrowser.org/`. This is an open source tool available for operating systems including Mac OS X, Windows, and Linux. It is a high quality tool used to create, edit, and design database with SQLite.

How to do it...

Perform the following steps to explore the data analysis:

1. Install SQLite once the download is complete.

2. Now, after installation, open SQLite and the window should look similar to the following screenshot:

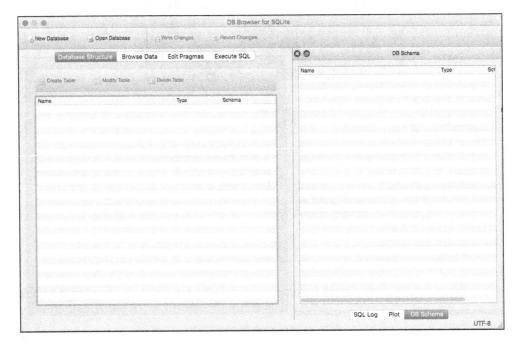

3. This tool provides a variety of options in which we create database files. Along with that it can also allow us to read the existing database files.

4. In order to read the existing database files, click on the **Open Database** button, which is at the top of the window.

5. You can navigate to the place where your .db file is located. Once located, select the file and click on **Open**.

6. Once the database is loaded successfully, it should look similar to the following screenshot:

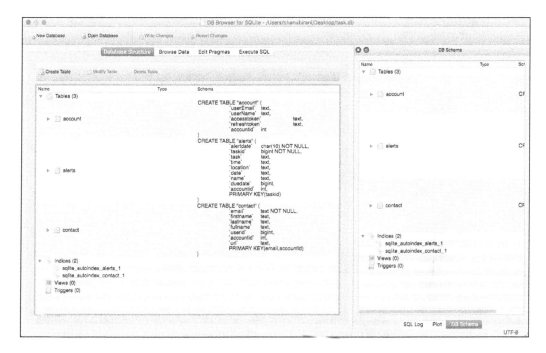

7. In the preceding screenshot, we can see that, by default, it has loaded the **Database Structure** tab, where we can see all the tables that exist in the database. In the current example, we have the **account**, **alerts**, and **contact** tables.

8. You can also create a new table by clicking on the **Create Table** button. Once you click on the button, you will get a popup, where you add the name of the table and other details about the fields. This should look similar to the following screenshot:

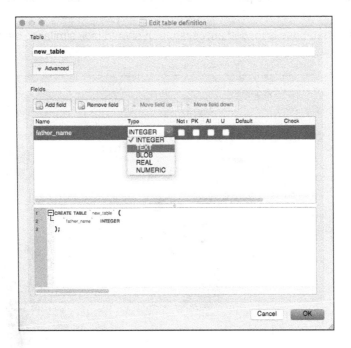

9. In the preceding screenshot, we are adding the table name `new_table` and field name `father_name`. For each field, we can set the type of value it will hold. Along with that, we can also pass various other information such as `Not null`, `Primary Key`, and so on.

10. SQLite browser also autogenerates a code for creating tables, which is executed at the end, once the **OK** button is clicked on.

11. You can also use the modify operation on the table by selecting any existing table and pressing the **Modify** button. This will give you a prepopulated UI, where you can edit any parameter or field.

12. You can perform various other operations such as adding, removing, and rearranging fields by clicking on the **Add field**, **Remove field**, and **Move field up** buttons respectively.

13. All your changes will be committed once you press the **OK** button. The UI should look similar to the following screenshot:

14. Now, we will see the way to read the data from the database. Click on the **Browse Data** tab in the window and you should be able to see the following screen:

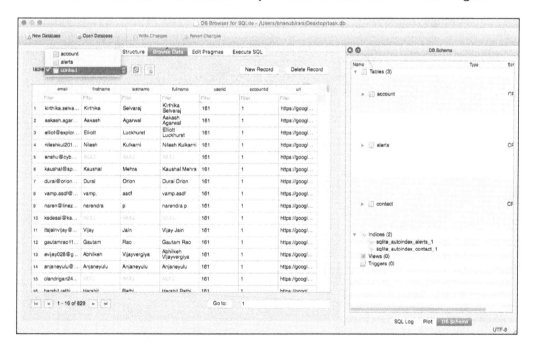

15. From the **Table** drop-down menu, you can select the table whose data needs to be explored. Once the table is selected, the relevant data is populated, which can be filtered again.

16. This also allows you to add a new record and delete a record using the **New Record** and **Delete Record** buttons respectively.

17. You can also apply a filter in order to show the matching results. For example, here, we are filtering the contact table with the `firstname` field. This filter will show you all the results that are exact matches to the text given filter area. This should look similar to the following screenshot:

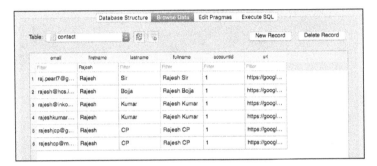

18. SQLite Browser also allows you to execute your custom SQL queries in order to filter results. For this, click on the **Execute SQL** tab. This tab should look like the following screenshot:

19. You can place your SQL query in the first section of the window. This window can be used as a debugging window in which we are exploring the database for the various outputs.

20. For example, we can start with the basic SQL query of fetching complete account table from the database. For this, please write the following query in the first section of the window:

```
select * from account
```

21. Now, click on the play button at the top to execute the query. This should show the output of the query in the second section of the window with logs in the third. The whole setup should look similar to the following screenshot:

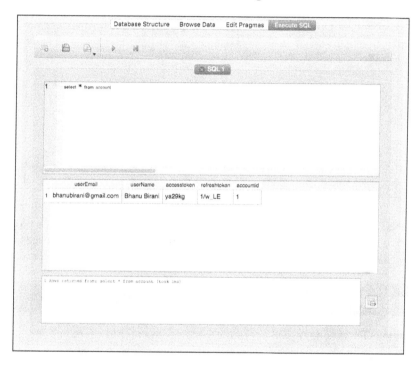

22. Here is one more complicated example of the query that you can execute in the SQLite browser window. The query is to filter the rows where `firstname` is `Rajesh` from the `contact` table. The query will be as follows:

```
select * from contact where firstname like 'Rajesh'
```

23. Click on the play button and you should be able to see the resultant rows in the second section of the window. This should look like the following screenshot:

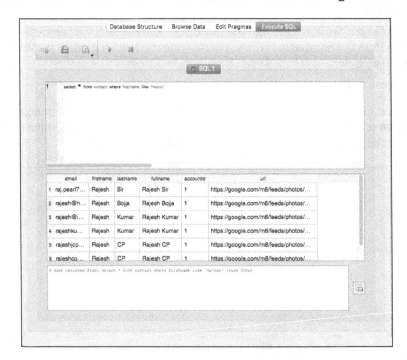

24. There can be various other creative ways to explore the data using this powerful open source tool. We have already seen how to extract data from Apple devices; this is a way to explore more details by looking at and extracting the data inside each file.

Jailbreaking iPhone devices

There are various limitations on iOS operating systems. In order to overcome those limitations, there is a way to jailbreak a device. In this section, we are going to explore the way to jailbreak an iDevice.

Getting ready

To jailbreak a device using OS X, you will need to download Pangu jailbreak software. You can download the Pangu software at `http://www.iphonehacks.com/download-pangu-jailbreak`. You need to download the latest version of Pangu for OS X.

How to do it...

Perform the following steps to jailbreak your iOS device using the Pangu software:

1. Connect your device to a computer using a USB cable.

2. Now, in order to proceed with the jailbreak, you will need to disable your device tracking and iCloud account. To disable **Find my iPhone**, navigate to **Settings | iCloud | Find my iPhone** and turn it off.

3. Now, you will need to disable the passcode and touch id from the phone; to do that, navigate to **Settings | Touch ID & Passcode and disable it**.

4. Then, turn on the airplane mode on the phone. This makes your device ready for the jailbreak.

5. Now, launch the Pangu application in the administrator mode.

6. The application will take some time to detect your device. Click on **Start** once the device is detected:

7. Now, click on the **Already backup** button to start the jailbreak process.

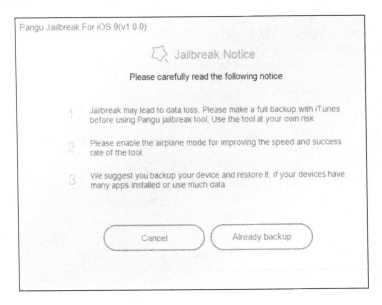

8. Once the jailbreak process is 55 percent completed, it should automatically reboot your device.

9. On completion of 65 percent of the process, it will prompt you to enable the airplane mode again after rebooting the device.

10. Once it is 75 percent completed, you will be prompted to unlock the device and run the Pangu app.

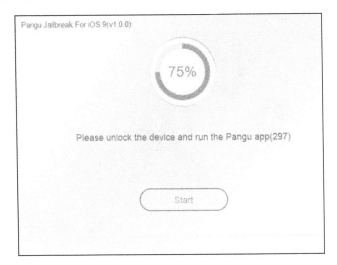

11. Now, you will be prompted to tap on the **Accept** button. This will explicitly give access to your iPhone Photos app.

12. Next, you just have to wait until the progress bar is completed. Once the process is finished, your device will reboot automatically.

13. After the successful completion of the process, your Pangu tool will show that your device is already jailbroken.

14. Now, turn off the airplane mode on your phone. You should also be able to see the Cydia app on your home screen; it will take a while when opening for the first time as it will be indexing the memory of your device.

15. Your jailbreaking process is completed.

Now, since you have jailbroken the iOS device, we can do a lot more with the phone that was restricted by Apple earlier. Jailbreaking offers more customization and access to various apps where we can share the device disk data with other mobile devices.

Index

Python
    download link 135

R

RAW file
    used, for saving data 3, 5
Recovery mode
    about 107
    launching, on iOS device 107, 108

S

SQLite browser
    download link 152
    exploring 151-158
SQLite database
    data, saving in 5-14

T

TestFlight
    app, distributing via 62-71
    setting up 61-71
tools
    decrypting 124-133
    encrypting 124-133

U

unique key
    generating 119

X

Xcode
    crash logs, receiving from device 98-100

Y

Yahoo Developer site
    reference link 48

Thank you for buying
iOS Forensics Cookbook

About Packt Publishing

Packt, pronounced 'packed', published its first book, *Mastering phpMyAdmin for Effective MySQL Management*, in April 2004, and subsequently continued to specialize in publishing highly focused books on specific technologies and solutions.

Our books and publications share the experiences of your fellow IT professionals in adapting and customizing today's systems, applications, and frameworks. Our solution-based books give you the knowledge and power to customize the software and technologies you're using to get the job done. Packt books are more specific and less general than the IT books you have seen in the past. Our unique business model allows us to bring you more focused information, giving you more of what you need to know, and less of what you don't.

Packt is a modern yet unique publishing company that focuses on producing quality, cutting-edge books for communities of developers, administrators, and newbies alike. For more information, please visit our website at www.packtpub.com.

About Packt Open Source

In 2010, Packt launched two new brands, Packt Open Source and Packt Enterprise, in order to continue its focus on specialization. This book is part of the Packt open source brand, home to books published on software built around open source licenses, and offering information to anybody from advanced developers to budding web designers. The Open Source brand also runs Packt's open source Royalty Scheme, by which Packt gives a royalty to each open source project about whose software a book is sold.

Writing for Packt

We welcome all inquiries from people who are interested in authoring. Book proposals should be sent to author@packtpub.com. If your book idea is still at an early stage and you would like to discuss it first before writing a formal book proposal, then please contact us; one of our commissioning editors will get in touch with you.

We're not just looking for published authors; if you have strong technical skills but no writing experience, our experienced editors can help you develop a writing career, or simply get some additional reward for your expertise.

Learning iOS Forensics

ISBN: 978-1-78355-351-8   Paperback: 220 pages

A practical hands-on guide to acquire and analyze iOS devices with the latest forensic techniques and tools

1. Perform logical, physical, and file system acquisition along with jailbreaking the device.

2. Get acquainted with various case studies on different forensic toolkits that can be used.

3. A step-by-step approach with plenty of examples to get you familiarized with digital forensics in iOS.

Practical Mobile Forensics

ISBN: 978-1-78328-831-1   Paperback: 328 pages

Dive into mobile forensics on iOS, Android, Windows, and BlackBerry devices with this action-packed, practical guide

1. Clear and concise explanations for forensic examinations of mobile devices.

2. Master the art of extracting data, recovering deleted data, bypassing screen locks, and much more.

3. The first and only guide covering practical mobile forensics on multiple platforms.

Please check **www.PacktPub.com** for information on our titles

Learning Android Forensics

ISBN: 978-1-78217-457-8   Paperback: 322 pages

A hands-on guide to Android forensics, from setting up the forensic workstation to analyzing key forensic artifacts

1. A professional, step-by-step approach to forensic analysis complete with key strategies and techniques.

2. Analyze the most popular Android applications using free and open source tools.

3. Learn forensically-sound core data extraction and recovery techniques.

Computer Forensics with FTK

ISBN: 978-1-78355-902-2   Paperback: 110 pages

Enhance your computer forensics knowledge through illustrations, tips, tricks, and practical real-world scenarios

1. Receive step-by-step guidance on conducting computer investigations.

2. Explore the functionality of FTK Imager and learn to use its features effectively.

3. Conduct increasingly challenging and more applicable digital investigations for generating effective evidence using the FTK platform.

Please check **www.PacktPub.com** for information on our titles

www.ingramcontent.com/pod-product-compliance
Lightning Source LLC
Chambersburg PA
CBHW060133060326
40690CB00018B/3865